D0947120

ELECTRICAL CRAFT PRINCIPLES

Volume 1

ELECTRICAL CRAFT PRINCIPLES

4th Edition

Volume 1

JOHN WHITFIELD

The Institution of Electrical Engineers

Published by: The Institution of Electrical Engineers, London,
United Kingdom

First published 1974
© 1974: Peter Peregrinus Ltd.

2nd Edition 1980
© 1980: Peter Peregrinus Ltd.

3rd Edition 1988
© 1988: Peter Peregrinus Ltd.

4th Edition 1995
© 1995: The Institution of Electrical Engineers

The Institution of Electrical Engineers,
Michael Faraday House,
Six Hills Way, Stevenage,
Herts. SG1 2AY, United Kingdom

British Library Cataloguing in Publication Data

A CIP catalogue record for this book
is available from the British Library

ISBN 0 85296 811 6

Printed in England by Short Run Press Ltd., Exeter

Contents

Preface to the Fourth Edition

This book has been on sale for the best part of 20 years through three editions, and the continuing demand for it has resulted in the need for this, the fourth edition.

Important changes have taken place since the appearance of the third edition in 1988. As well as the introduction of the 16th Edition of the IEE Wiring Regulations, this extremely important set of wiring rules has become a British Standard, *BS 7671* (Requirements for Electrical Installations). There have also been changes in the syllabuses of the examining bodies responsible for the certification of electrical craftsmen, as well as in the pattern of education and training which they are required to follow.

The increased 'hands-on' nature of the training is to be welcomed, as is the widening of syllabuses to include more electronics. Multiple-choice questions are now in wide use, and a selection of this type of test has been provided in each Chapter of this book. I must admit to having doubts about presenting the reader with faulty information in the wrong answers to the multiple-choice questions, but am sure that a sensible approach, including the analysis of each wrongly answered question, will ensure that no harm results. I still feel very strongly that the best method of mastering material of this kind is to work through numerous examples. To this end, the present volume contains 98 fully worked examples, 304 exercises and 231 multiple choice exercises, all with answers.

When originally written, the first volume of this book covered Part I of the craft syllabuses. Changes have meant that this fourth edition now covers most of Part II as well. No attempt has been made to deal with the practical-training aspects, which require a much more detailed analysis than is possible in a single volume.

The book will be found suitable as the main textbook or for providing essential background learning for a wide range of courses, which include:

City & Guilds 2320 Electrical and Electronic Craft Studies
City & Guilds 2360 Electrical Installation Work
City & Guilds 2210 Telecommunication and Electronics Mechanics
City & Guilds 2380 16th Edition of the IEE Wiring Regulations
City & Guilds 2710 Telecommunications Technicians
City & Guilds 2900 Essential Electricity

The author believes that the material included in this book, in addition to being suitable for craft students, is suitable for much of the early study in the BTEC courses leading to qualifications in electrical and electronic engineering.

John Whitfield
Norwich
January 1995

Acknowledgments

None of the work in this book can be claimed to be original, and the author acknowledges with thanks the numerous teachers, students and organisations who have contributed to his understanding of the subject. Unfortunately, they are too numerous to mention, but the author would like to acknowledge:

- The Institution of Electrical Engineers for permission to quote from its Wiring Regulations
- N. Hiller for his continuing support and help
- His wife for her continued help and forbearance

and

- Nigel Thomas of Saft Nife Ltd
- J. I. Greatorex of Duracell Ltd
- A. I. Harrison of Chloride Industrial Batteries Ltd
- R. N. Green of BICC Cables Ltd

for assistance with technical information and diagrams.

Symbols and abbreviations

1 Terms

Term	Symbol or abbreviation
alternating current	AC
area	A or a
area, cross-sectional	CSA
capacitance	C
charge (electric)	Q
current	
steady or RMS value	I
instantaneous value	i
maximum value	I_m
average value	I_{av}
direct current	DC
distance	d
electromotive force	EMF
steady or RMS value	E
instantaneous value	e
energy	W
force	F
frequency	f
inductance, self	L
length	l
line current	I_L
line voltage	U_L
magnetic flux	Φ (phi)
magnetic flux density	B
magnetising force	H
magnetomotive force	MMF
mass	m
mechanical advantage	MA
permeability of free	
space	μ_0 (mu)
permeability, relative	μ_r (mu)
phase angle	ϕ (phi)
phase current	I_P

Term	Symbol or abbreviation
phase voltage	U_P
potential difference	PD
steady or RMS value	U
instantaneous value	u
maximum value	U_m
average value	U_{av}
power	P
primary current	I_1
primary EMF	E_1
primary turns	N_1
primary voltage	U_1
quantity of electricity	Q
reactance, capacitive	X_c
reactance, inductive	X_L
resistance	R
resistivity	ρ (rho)
revolutions per minute	r/min
revolutions per second	r/s
root mean square	RMS
rotational velocity	ω (omega)
secondary current	I_2
secondary EMF	E_2
secondary turns	N_2
secondary voltage	U_2
speed (r/min)	N
speed (r/s)	n
temperature coefficient	
of resistance	α (alpha)
time	t
time, periodic	T
torque	T
velocity ratio	VR
work	W

2 Units

Unit	Symbol	Unit of
ampere	A	electric current
ampere-turn	At	magnetomotive force
coulomb	C	electric charge or quantity
cycle per second (or hertz)	c/s (or Hz)	frequency
degree Celsius	°C	temperature
farad	F	capacitance
henry	H	inductance
hertz	Hz	frequency
joule	J	work or energy
kilogram	kg	mass
kilowatt	kW	power
kilowatt-hour	kWh	energy
metre	m	length
millimetre	mm	length
newton	N	force
newton metre	Nm	torque
metre newton (joule)	mN	work and energy
ohm	Ω (omega)	resistance
radians per second	rad/s	rotational velocity
second	s	time
tesla	T	magnetic flux density
volt	V	PD and EMF
watt	W	power
weber	Wb	magnetic flux
weber per square metre	Wb/m^2	magnetic flux density

3 Multiples and submultiples

Prefix	Symbol	Meaning	
meg- or mega-	M	one million times	or $\times 10^6$
kil- or kilo-	k	one thousand times	or $\times 10^3$
milli-	m	one-thousandth of	or $\times 10^{-3}$
micro-	μ (mu)	one-millionth of	or $\times 10^{-6}$
nano-	n	one-thousand-millionth of	or $\times 10^{-9}$
pico-	p	one-million-millionth of	or $\times 10^{-12}$

SI units

If we are to measure physical, mechanical and electrical quantities, we must use a system of units for the purpose. All units, no matter how complex, are based on a number of basic units. The system of units adopted in much of the world is the SI (an abbreviation of Système International d'Unités):

Physical quantity	Name of unit	Symbol
length	metre	m
mass	kilogram	kg
time	second	s
electric current	ampere	A
temperature	kelvin	K
luminous intensity	candela	cd

Imperial units (such as the inch or the pound) are not recognised in this volume.

Multiple and submultiple units

There are many examples in practical electrical engineering where the basic units are of inconvenient size.

Multiple units are larger than the basic units.

The prefix **meg** or **mega** (symbol M) means **one million times**.

For instance,   1 megavolt (1 MV) = 1 000 000 volts

and   1 megohm (1 MΩ) = 1 000 000 ohms

The prefix **kil** or **kilo** (symbol k) means **one thousand times**.

For instance,   1 kilovolt (1 kV) = 1 000 volts.

Submultiple units are smaller and are decimal fractions of the basic units.

The prefix **milli** (symbol m) means **one-thousandth of**.

For instance,   1 milliampere (1 mA) = $\dfrac{1}{1000}$ ampere

The prefix **micro** (symbol μ, the Greek letter 'mu') means **one-millionth of**.

For instance,   1 microhm (1 $\mu\Omega$) = $\dfrac{1}{1\ 000\ 000}$ ohm

The prefix **nano** (symbol n) means **one-thousand-millionth of**.

For instance,   1 nanosecond (1 ns) = $\dfrac{1}{1\ 000\ 000\ 000}$ second

The prefix **pico** (symbol p) means **one-million-millionth of**.

For instance,   1 picovolt (1 pV) = $\dfrac{1}{1\ 000\ 000\ 000\ 000}$ volt

Two words of warning are necessary concerning the application of these extremely useful prefixes. First, note the difference between the symbols M and m. The ratio is one thousand million! Secondly, always convert a value into its basic unit before using it in an equation. If this is done, any unknown value in the equation can be found in terms of its basic unit.

Chapter 1

Basic electric units and circuits

1.1 Simple electron theory

Electricity in the form of lightning must have been apparent to man from his earliest cave-dwelling days. The use of electricity has increased many times during the last 80 years, and with this increased use has come a greater awareness of the nature of electricity. Present-day theories are based on the theory of atomic structure, although our knowledge is still far from complete.

The atom

All matter is composed of atoms, which often arrange themselves into groups called **molecules**. An atom is so very small that our minds are unable to appreciate what vast numbers of them make up even a very small piece of material. Eight million typical atoms, placed end to end, would stretch for about 1 mm.

The atom itself is not solid, but is composed of even smaller particles separated by space. At the centre of each atom is the **nucleus**, which is made up of various particles, including **protons**. These protons are said to have a **positive charge**. The **electrons**, which complete the atom, are in a constant state of motion, circling the nucleus in the same way as a satellite circles the Earth. Each electron has a **negative charge**. Atoms of different materials differ from one another by having different numbers of electrons but, in the complete state, each atom has equal numbers of protons and electrons, so that positive and negative charges cancel out to leave the atom electrically neutral. The atoms in solids and liquids are much closer together than they are in gases, and in solids they are held in a definite pattern for a given material.

Where there are more than two electrons in an atom, their paths of motion are in **shells**. The electron paths are, in fact, elliptical rather than circular, and take all directions around the nucleus. The atom is thus

1

three-dimensional, and not two-dimensional as indicated in the simplified form of Figure 1.1, which shows a simple representation of a copper atom, which has 29 electrons and 29 protons. The electron in the outer shell is weakly held in position, and often breaks free, moving at random among the other copper atoms. An atom that has lost an electron in this way is left with an overall positive charge, since it has a positive proton in excess of those required to balance the effect of its negative electrons. Such an incomplete atom is often called a **positive ion**.

The movement of free electrons in a conductor depends on the laws of electric charge, which are that

(*a*) like charges repel, and
(*b*) unlike charges attract.

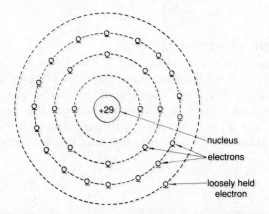

Figure 1.1   Simplified representation of copper atom

Electric current

Figure 1.2*a* represents a block of conducting material, containing free electrons moving at random among positive ions. If a battery were connected across the block as shown in Figure 1.2*b*, free electrons close to the positive plate will be attracted to it, since unlike charges attract. Free electrons near the negative plate will be repelled from it, and a steady drift of electrons will take place through the material from the negative battery terminal to the positive battery terminal. For each electron entering the positive terminal, one will be ejected from the negative terminal, so that the number of electrons in the material remains constant. Since the atoms that have become positive ions are unable to move in a solid, they do not drift to the negative terminal.

The rate of movement of electrons through the solid is very low, but, since free electrons throughout the material start to drift immediately the battery is connected, there is very little delay in the demonstration of the effects that occur as a result of this movement.

Figure 1.2   Movement of electrons
  a Random movement of electrons in conductor
  b Drift of electrons towards positive plate when battery is connected

The drift of electrons *is* the electric current; so, to some extent, we have been able to answer the question 'What is electricity?' However, we have no clear understanding of the nature of an electron, so our knowledge is far from complete.

Electrons which enter the battery through the positive plate are passed through it, and are ejected from the negative plate into the conductor. Thus the electrons circulate, but must have a continuous conducting path, or **closed circuit**, in which to do so. If the circuit is broken, the drift of electrons will cease immediately.

Current direction

The knowledge that an electric current consists of a drift of electrons is of comparatively recent origin. Long before this theory was put forward, electric current was thought of as due to an 'electric fluid' which flowed in conductors from the positive plate of a battery to the negative. This direction of current, called **conventional current direction**, was thought to be correct for many years, so that many rules were based on it. We now know that this assumed current direction is incorrect and that current in a solid actually consists of an electron drift in the opposite direction. Despite this, we still continue, by convention, to indicate current direction, external to the source, as being from the positive to the negative terminal. In most of the applications we shall consider, the actual direction of current does not affect the performance of equipment; because of this, we shall continue to use conventional current direction.

The directions of electron drift and conventional current are shown in Figure 1.3.

Figure 1.3   Circuit to illustrate heating, magnetic and chemical effects of electric current

1.2   Electric charge and unit of current

If we wish to measure a length we do so with a rule which is marked off in specific units of length. Since an electric current is invisible, we must use special instruments for measuring it, most of these instruments depending for their operation on the magnetic field set up by a current in a conductor. These instruments are described in Section 1.9.

The first unit we shall consider is the unit of electric charge or quantity of electricity. It may seem that the electron could be used, but it is far too small for practical purposes. The unit used is the **coulomb** (symbol C), which is very much larger than the electron, the charge of over six million million million electrons equalling that of one coulomb. If a body has a surplus of electrons it is said to be **negatively charged**, whereas if it has a shortage of electrons it is **positively charged**. Both these amounts of charge could be measured in coulombs.

A newcomer to the theory of electricity may be struck by the curious names applied to units. Most units are named after great scientists, like the unit of electric charge, the coulomb, which is named after Charles Coulomb, a French physicist (1736–1806).

If the drift of electrons in a conductor takes place at the rate of one coulomb per second, the resulting current is said to be a current of one **ampere** (symbol A). Thus, a current of one ampere indicates that charge is being transferred along the conductor at the rate of one coulomb per second; hence

$$Q = It$$

where Q = charge transferred in coulombs
    I = current in amperes
    t = time during which the current flows in seconds

Example 1.1

If a total charge of 500 C is to be transferred in 20 s, what is the current flow?

$$Q = It \text{ and thus } I = \frac{Q}{t}$$

Therefore $$I = \frac{500}{20} \text{ amperes} = 25 \text{ A}$$

Example 1.2

A current of $12\cdot5$ A passes for 2 min. What quantity of electricity is transferred?

$$Q = It$$
$$= 12\cdot5 \times 2 \times 60 \text{ coulombs}$$
$$= 1500 \text{ C}$$

Example 1.3

A current of $0\cdot15$ A must transfer a charge of 450 C. For how long must the current pass?

$$Q = It \text{ so } t = \frac{Q}{I}$$
$$t = \frac{450}{0\cdot15} \text{ seconds}$$
$$= 3000 \text{ s or } 50 \text{ min}$$

1.3 Effects of electric current

Electrons are far too small to be seen even with the best available microscopes, and the detection of current would be impossible if it did not produce effects that are more easily detected. There are many such effects, but the three most important are **heating**, **chemical** and **magnetic**.

When current flows in a wire, heat is generated. The amount of heat produced in this way depends on a number of factors, which will be considered later, but can be controlled by the variation of current, of conductor material and of conductor dimensions. In this way, the conductor can be made red or white hot as with an electric fire or filament lamp, or can be made to carry current and remain reasonably cool as with an electric cable.

When current passes through chemical solutions, it can cause basic chemical changes to take place in them. Examples of this are the battery of cells, and electroplating. Some of these chemical effects will be further considered in Chapter 8.

A current flowing in a coil gives rise to a magnetic field, and this principle is the basis of many electrical devices such as the motor, the relay and the bell. The magnetic effect is the subject of Chapter 6. Figure 1.3 shows a circuit in which the same current passes in turn through a filament lamp (heating effect), an electromagnet (magnetic effect) and a lead–acid cell (chemical effect). The heating and magnetic effects will be apparent owing to the heating of the lamp filament and the attraction of the iron armature. The chemical effect is demonstrated if the changeover switch is operated, when energy stored in the cell will cause the small filament lamp to glow.

Table 1.1 lists some of the common devices relying on these three electrical effects.

Table 1.1: Some devices relying on effects of electric current

Magnetic effect	Heating effect	Chemical effect
Relay	Filament lamp	Cells and batteries
Bell	Electric heater	Electroplating
Contactor	Electric cooker	Fuel cell
Telephone	Electric iron	
Motor	Electric kettle	
Generator	Television tube	
Transformer	Fuse	
Circuit breaker	Circuit breaker	
Tape recorder	Welder	
Ammeter	Furnace	
Voltmeter		

1.4 Electric conductors and insulators

Electric conductors

It has been stated that an electric current is the drift of free electrons in a solid. It follows that, for a material to be capable of carrying current, the atoms of which it is composed will have loosely held electrons, which become detached at normal temperatures or can be detached by the application of an electric charge. Such materials are celled **electric conductors**. A list of conductors, with remarks on their properties and uses, is given in Table 1.2.

Silver is the best electric conductor, but its high cost and poor physical properties prevent its use as a cable material. Copper is next in conducting properties to silver. Its malleability (the ease with which it can be beaten

Table 1.2: Electric conductors

Material	Properties	Application
Aluminium	Low cost and weight	Power cables
Brass	Easily machined; resists corrosion	Terminals, plug pins
Carbon	Hard; low friction with other metals	Machine brushes
Chromium	Hard; resists corrosion	Heating elements (with nickel)
Copper	Very good conductor; soft and easily drawn into wires; capable of hardening	All cables, busbars
Gold	Expensive; does not corrode	Plating on contacts
Iron and steel	Common metal	Conduits, trunking, fuseboard cases etc. (protective conductor)
Lead	Does not corrode; bends easily	Power cable sheaths (protective conductor)
Mercury	Liquid at normal temperatures; vaporises readily	Special contacts, discharge lamps
Nickel	Hard; resists corrosion	Heating elements (with chromium)
Silver	Expensive; the best conductor	Fine instrument wires, plating on contacts
Sodium	Vaporises readily	Discharge lamps
Tin	Resists attack by sulphur	Coating on copper cables insulated with vulcanised rubber
Tungsten	Easily drawn into fine wires	Lamp filaments

into shape) and its ductibility (the ease with which it can be drawn into strands) make it the natural choice as a conductor for cables; many heavy supply cables, and almost all wiring cables, have copper conductors.

Aluminium is a poorer conductor than copper, but it is lighter. As copper prices have increased in recent years, aluminium prices have remained stable, so aluminium is a direct competitor to copper for power cables. Since aluminium is not as flexible as copper, cannot be drawn into such fine wires and poses connection problems owing to rapid surface corrosion, the IEE Wiring Regulations (*BS 7671*) forbid its use in the form of small wiring cables. Such cables, however, are made and used.

Electric insulators

If materials are composed of atoms which have all their electrons tightly bound to them, there will be no free electrons available to form an electric current, and none can flow. Such materials are called **electric insulators**.

There are very many types of insulating material, but a few of those in common use in the electrical industry are listed in Table 1.3.

Table 1.3 is far from complete. For example, many new types of plastics have been developed as cable insulation, each having special properties. Polyvinyl chloride (PVC) is the most frequently used insulating and sheathing material for internal normal-temperature applications. Polychloroprene (PCP) has particularly good weather-resisting properties, as has chlorosulphonated polyethelyne (CSP), which also has increased resistance to physical damage. These are but three examples of the numerous insulating materials now available to the engineer and craftsman.

No material is a perfect insulator, and all will pass a small 'leakage current'. This leakage is usually so small compared with the operating currents of the equipments that it may be ignored.

Conductors and insulators

An electric cable is a very good example of the application of conductors and insulators. Figure 1.4 shows a typical twin housewiring cable with protective conductor, the copper current-carrying conductors being insulated with PVC. An overall sheath of PVC keeps the conductors together, and protects them from damage and dampness. The conductors are made of

Table 1.3: Electric insulators

Material	Properties	Application
Rubber Flexible plastics }	Flexible; life affected by high temperature	Cable insulation (small and medium sizes)
Crosslinked polyethylyne (XLPE)	Emits little smoke or fumes when burning	Cable insulation (medium and large sizes)
Impregnated paper Varnished cambric }	Rather stiff, but unaffected by moderate temperatures; hygroscopic	Cable insulation (medium and large sizes)
Magnesium oxide (mineral insulation)	Powder; requires containing sheath; not affected by very high temperatures; very hygroscopic	Mineral-insulated cables
Mica	Insulation not affected by high temperatures	Kettle elements, toaster elements etc.
Asbestos Glass fibre }	Reasonably flexible; not affected by high temperatures	Cable insulation in cookers, fires etc.
Glass	Rigid, easily cleaned	Overhead-line insulators
Porcelain	Hard and brittle; easily cleaned	Fuse carriers, overhead-line insulators etc.
Rigid plastics	Not as expensive as porcelain, and less brittle	Fuse carriers, switches sockets, plugs etc.

Figure 1.4   Twin-and-protective-conductor PVC-insulated PVC-sheathed wiring cable

copper, which is softened by annealing to make it flexible. The flexibility is sometimes further improved by using stranded, instead of solid, conductors.

Semiconductors

Semiconductors have electrical properties lying between those of conductors and insulators. They occupy a very important place in such devices as rectifiers and transistors, which will be considered in Chapter 15.

1.5   Electrical energy, work and power

Before we can go on to consider the electrical force that results in electron drift in a conductor, we must look at the units used for measuring work and power. Fuller consideration of electrical energy, work and power is given in Chapter 5.

Energy and work

Energy and work are interchangeable, energy being used to do work. Both are measured in terms of force and distance. The SI unit of force is the **newton** (N), and the unit of distance or length is the **metre** (m). If a force moves through a distance, work is done and energy is used.

energy used = work done = distance moved (m) × force for movement (N)

Since the distance moved is in metres and the force required for the movement is in newtons, the unit is the metre-newton, more commonly called the **joule** (J).

Example 1.4

A force of 2000 N is required to lift a machine. How much energy is required to lift the machine through 3 m?

$$\text{work done} = \text{distance} \times \text{force}$$

$$= 3 \times 2000 \text{ joules}$$

$$= 6000 \text{ J}$$

This work is mechanical, but we shall see that work can also be electrical.

Power

Power is the rate of doing work or of using energy. For instance, an electrician can cut a hole in a steel plate using a hand drill or an electric drill. With both, the effective work done will be the same, but the electric drill will cut the hole more quickly because its power is greater.

It follows that

$$\text{power} = \frac{\text{work or energy}}{\text{time}}$$

The SI unit of power is the **watt** (W), which is a rate of doing work of one joule per second.

$$\text{watts} = \frac{\text{joules}}{\text{seconds}}$$

Similarly, if we know how much power is being used and the time for which it is used, we can find the total energy used:

$$\text{work or energy} = \text{power} \times \text{time}$$

$$\text{or joules} = \text{watts} \times \text{seconds}$$

Example 1.5

A works truck requires a force of 180 N to move it. How much work is done if the truck is moved 20 m, and what average power is employed if the movement takes 40 s?

$$\text{work} = \text{distance} \times \text{force}$$

$$= 20 \times 180 \text{ joules}$$

$$= 3600 \text{ J}$$

$$\text{average power} = \frac{\text{work}}{\text{time}}$$

$$= \frac{3600}{40} \text{ watts}$$

$$= 90 \text{ W}$$

1.6 Electromotive force and potential difference

When an electric current flows, energy is dissipated. Since energy cannot be created, it must be provided by the device used for circulating the current. This device may be chemical, such as a battery; mechanical, such as a generator; or it may have one of a number of other forms. Many years ago, electricity was thought to be a fluid which circulated as the result of a force, and thus the term **electromotive force** (EMF), symbol E, came into use. The EMF is measured in terms of the number of joules of work necessary to move one coulomb of electricity round the circuit, and thus has the unit joules/coulomb. This unit is referred to as the **volt** (symbol V) so that

$$1 \text{ volt} = 1 \text{ joule/coulomb}$$

Example 1.6

A battery with an EMF of 6 V gives a current of 5 A round a circuit for five minutes. How much energy is provided in this time?

$$\text{total charge transferred } Q = It = 5 \times 5 \times 60 \text{ C} = 1500 \text{ C}$$

$$\text{total energy supplied} = (\text{joules/coulomb}) \times \text{coulombs}$$

$$= \text{volts} \times \text{coulombs}$$

$$= 6 \times 1500 \text{ joules} = 9000 \text{ J}$$

In example 1.6, each coulomb of electricity contained six joules of potential energy on leaving the battery. This energy was dissipated on the journey round the circuit, so that the same charge would possess no energy on its return to the battery. The amount of energy expended by one coulomb in its passage between any two points in a circuit is known as the **potential difference** (PD) between those points, and is measured in joules/coulomb, or volts. The international symbol for supply voltage or voltage drop has changed from V to U, the symbol for the unit of voltage (the volt) remaining as V.

A convenient definition of the volt is therefore that it is equal to the difference in potential between two points if one joule of energy is required to transfer one coulomb of electricity between them.

Example 1.7

How much electrical energy is converted into heat each minute by an immersion heater which takes 13 A from a 240 V supply?

$$\text{energy given up by each coulomb} = 240 \text{ J}$$

$$\text{quantity of electricity flow per minute} = It = 13 \times 60 \text{ coulombs} = 780 \text{ C}$$

$$\text{therefore energy converted in one minute} = (\text{joules/coulomb}) \times \text{coulombs}$$

$$= 240 \times 780 \text{ joules} = 187\ 200 \text{ J}$$

1.7   Resistance: Ohm's law

For a metallic conductor which is kept at a constant temperature, it is found that the ratio

$$\frac{\text{potential difference across conductor (volts)}}{\text{resulting current in conductor (amperes)}}$$

is constant, and this ratio is known as the **resistance** (symbol R) of the conductor. This important relationship was first verified by Dr. G. S. Ohm, and is often referred to as 'Ohm's law'. The unit of resistance is the **ohm** (Greek symbol Ω, 'omega'). A conductor has a resistance of one ohm if the PD across its ends is one volt when it carries a current of one ampere. A device intended to have a resistance is called a **resistor**.

The relationship expressed by Ohm's law, which is of fundamental importance in electrical engineering, can be simply written as a formula

$$U = I \times R$$

The subject of the formula can be changed to give

$$I = \frac{U}{R} \text{ or } R = \frac{U}{I}$$

Example 1.8

An electrical heater used on a 240 V supply carries a current of 12 A. What is its resistance?

$$R = \frac{U}{I}$$

$$= \frac{240}{12} \text{ ohms}$$

$$= 20 \ \Omega$$

Example 1.9

The insulation resistance between two cables is two million ohms (2 MΩ). What leakage current will flow if a PD of 400 V exists between them?

$$I = \frac{U}{R}$$

$$= \frac{400}{2\ 000\ 000} \text{ amperes}$$

$$= 0 \cdot 000\ 2 \text{ A or } 0 \cdot 2 \text{ mA}$$

Example 1.10

What PD exists across an earth-continuity conductor of resistance $1 \cdot 2 \ \Omega$ when a current of 25 A flows through it?

$$U = IR$$

$$= 25 \times 1 \cdot 2 \text{ volts}$$

$$= 30 \text{ V}$$

1.8 Electrical circuit

For an electromotive force to drive electrons, it must be applied to a closed circuit. In practice, the circuit is likely to consist of a piece of apparatus connected to the source of EMF by means of cables, which complete the circuit. Figure 1.5 shows a simple circuit consisting of a source of EMF, and a resistor. A switch, shown closed in Figure 1.5*a* is included in the circuit. Since the circuit is completed, the EMF of the supply will provide a current, its value depending on the voltage of the supply and on the resistance of the circuit.

If the load has high resistance, the current will be small; if the load resistance is low, the current will be greater.

If the switch is opened, as in Figure 1.5*b*, the gap between the opened contacts introduces a nearly infinite resistance into the circuit, so that the current falls to zero. We say that opening the switch has 'broken' the circuit.

Figure 1.5 EMF in a circuit
a Closed circuit
b Open circuit

Although there are some devices, including electronic visual display units (VDUs) and fluorescent lamps, which do not offer a complete metallic path for current, the majority of circuits are made up entirely of such conductors. If the conducting path is interrupted, the current ceases.

In circuits where high voltages are present, opening a switch may not break the circuit, the current continuing through the air between the contacts. The air carrying the current glows brightly, gives off heat and is called an **arc**. The majority of switches produce an arc when opened, but in most cases the arc disappears within a fraction of a second, and the circuit is broken before the heat from the arc can damage the switch or its surroundings.

1.9 Ammeters and voltmeters

Although the presence of an electric current may produce effects which can be detected by the human senses, such effects are seldom useful as an indication of the value of the current. For instance, when a filament lamp glows, it is clearly carrying current, although we are unlikely to be able to judge the value. However, when the current in the lamp is reduced

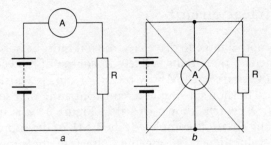

Figure 1.6   Correct (a) and incorrect (b) connection of ammeter

Figure 1.7   Correct (a) and incorrect (b) connection of voltmeter

to about one-third of its normal value, the lamp ceases to glow.

Instruments for direct measurement of electric current are called **ammeters**, and will be considered in detail in the second volume. The principles of two types of instrument are discussed in Sections 7.8 and 11.6. Ammeters have low resistance, and are connected so that the current to be measured passes through them. Figure 1.6 shows correct (*a*) and incorrect (*b*) ammeter connections. The ammeter will be damaged if incorrectly connected.

Figure 1.6 makes it clear that the circuit symbol for an ammeter is a circle containing the letter A.

A **voltmeter** is another measuring instrument, used to indicate the potential difference between its two connections. To give an indication, the voltmeter must be connected across the device or circuit whose PD is to be indicated. Figure 1.7 gives correct (*a*) and incorrect (*b*) voltmeter connections, and shows the voltmeter symbol as a circle containing the letter V. If connected incorrectly, the voltmeter is unlikely to be damaged, but, since it has high resistance, it will prevent the correct functioning of the circuit. The source of supply in the circuits of Figures 1.6 and 1.7 is a battery of cells.

1.10   Series circuits

When a number of resistors are connected together end to end, so that there is only one path for current through them, they are said to be connected **in series**. An electrical appliance is connected in series with the cables feeding it; and, since the total current will depend on the resistance of the circuit as well as on the voltage applied to it, it is important to be able to calculate the resistance of the complete circuit if we know the values of the individual resistors connected in it. Figure 1.8 shows three resistors, of values R_1, R_2 and R_3, respectively, connected in series across a supply of U volts. Let us assume that the resulting current is I amperes. If the total circuit resistance is R ohms,

$$R = \frac{U}{I}$$

Now let the PD across each of the three resistors be U_1, U_2 and U_3 volts, respectively.

Figure 1.8   Resistors connected in series

Then $\qquad\qquad U_1 = IR_1, \ U_2 = IR_2 \ \text{and} \ U_3 = IR_3$

but $\qquad\qquad U = U_1 + U_2 + U_3$

$$= IR_1 + IR_2 + IR_3$$

$$= I(R_1 + R_2 + R_3)$$

$$\frac{U}{I} = R_1 + R_2 + R_3$$

but $\qquad\qquad \dfrac{U}{I} = R, \text{ so that}$

$$R = R_1 + R_2 + R_3$$

Thus the total resistance of any number of resistors connected together in series can be found by adding the values of the individual resistors, which must all be expressed in the same unit.

Example 1.11

Resistors of 50 Ω and 70 Ω are connected in series to a 240 V supply. Calculate (*a*) the total resistance of the circuit (*b*) the current and (*c*) the PD across each resistor.

(*a*) $R = R_1 + R_2 = 50 + 70 \text{ ohms} = 120 \ \Omega$

(b) $I = \dfrac{U}{R} = \dfrac{240}{120}$ amperes = 2 A

(c) $U_1 = IR_1 = 2 \times 50$ volts = 100 V

   $U_2 = IR_2 = 2 \times 70$ volts = 140 V

Note: Supply voltage $U = U_1 + U_2 = 100 + 140$ volts = 240 V

   Note that, for a series circuit,

(i)   the same current flows in all resistors
(ii) the PD across each resistor is proportional to its resistance
(iii) the sum of the PDs across individual resistors is equal to the supply voltage.

Example 1.12

An electric heater consists of an element of resistance 23·8 Ω and is fed from a 240 V supply by a 2-core cable of unknown resistance. If the current is 10 A, calculate the total resistance of the cable.

   Figure 1.9a shows the heater connected to the supply through a two-conductor cable. Figure 1.9b is an equivalent circuit in which conductors are assumed to have no resistance, the actual resistance of the conductors being replaced by the resistor R. There are two methods of solution.

Method 1

Total resistance = $\dfrac{U}{I} = \dfrac{240}{10}$ ohms = 24 Ω

But total resistance = element resistance + conductor resistance
Therefore conductor resistance = total resistance – element resistance
Conductor resistance = 24 – 23·8 ohms = 0·2 Ω

Method 2

PD across element = current × element resistance
         = 10 × 23·8 volts = 238 V

Figure 1.9   Circuit diagrams for Example 1.12

Supply voltage = PD across element + PD across conductors
Therefore PD across conductors = supply voltage – PD across element
$$= 240 - 238 \text{ volts} = 2 \text{ V}$$

$$\text{Conductor resistance} = \frac{\text{conductor PD}}{\text{current}}$$

$$= \frac{2}{10} \text{ ohms} = 0 \cdot 2 \text{ } \Omega$$

1.11   Parallel circuits

When each one of a number of resistors is connected between the same two points, they are said to be connected **in parallel**. In this form of connection, the total current divides, part of it flowing in each resistor. Since all the resistors are connected across the same two points, the PD across each one is the same. Figure 1.10 shows resistors of values R_1, R_2 and R_3, respectively, connected in parallel to a supply of U volts. The currents in the resistors are I_1, I_2 and I_3, respectively, and the total current is I.

The total current divides itself among the resistors, so that

$$I = I_1 + I_2 + I_3$$

But, from Ohm's law,

$$I_1 = \frac{U}{R_1} \qquad I_2 = \frac{U}{R_2} \qquad I_3 = \frac{U}{R_3}$$

$$I = \frac{U}{R_1} + \frac{U}{R_2} + \frac{U}{R_3} = U\left(\frac{1}{R_1} + \frac{1}{R_2} + \frac{1}{R_3}\right)$$

$$\frac{I}{U} = \frac{1}{R_1} + \frac{1}{R_2} + \frac{1}{R_3}$$

If the equivalent resistance of the parallel circuit is R, $R = U/I$.

Figure 1.10   Resistors connected in parallel

Therefore

$$\frac{I}{U} = \frac{1}{R} = \frac{1}{R_1} + \frac{1}{R_2} + \frac{1}{R_3}$$

The value $1/R$ is called the **reciprocal** of R. We can thus sum up the expression by saying that the reciprocal of the equivalent resistance of a parallel circuit is equal to the sum of the reciprocals of the resistances of the individual resistors.

Note that, for a parallel circuit,

(*a*) the same PD occurs across all resistors
(*b*) the current in each resistor is inversely proportional to its resistance
(*c*) the sum of the currents in the individual resistors is equal to the supply current.

As for the series circuit, the resistances must all be expressed in the same unit before using them in the formula.

Example 1.13

Calculate the total resistance of three parallel-connected resistors of 6 Ω, 30 Ω and 10 Ω, respectively.

$$\frac{1}{R} = \frac{1}{R_1} + \frac{1}{R_2} + \frac{1}{R_3}$$

$$\frac{1}{R} = \frac{1}{6} + \frac{1}{30} + \frac{1}{10} = \frac{5+1+3}{30} = \frac{9}{30}$$

$$R = \frac{1}{1/R} = \frac{30}{9} \text{ ohms} = 3\cdot33 \text{ } \Omega$$

It can be seen from this result that the equivalent resistance of any group of parallel-connected resistors is lower than that of the lowest-valued resistor in the group. If a number of equal-value resistors are connected in parallel, the equivalent resistance will be the value of one resistor divided by the number of resistors. For example, ten 10 Ω resistors in parallel have an equivalent resistance of 1 Ω.

In the example above, it has been simple to find the lowest common denominator (30) into which all three resistor values divide exactly. Had the resistors had less convenient values, the process could have been very difficult and it is sensible to use the calculator for the process.

To do this, we need to identify a number of dedicated keys. These are the reciprocal key (usually marked '1/x'), the accumulating memory key (usually marked 'SUM') and the memory recall key (usually marked 'RCL'). To solve example 1.13 using the calculator proceed as follows. First press the clear button (marked 'AC') to make sure that the memories are empty.

Action	Display shows	Explanation
Enter 6	6	enters R_6
Press 1/x	0·1666666	calculates $1/R_6$

Press SUM	0·1666666	sends display to memory
Enter 30	30	enters R_{30}
Press 1/x	0·0333333	calculates $1/R_{30}$
Press SUM	0·0333333	adds display to memory
Enter 10	10	enters R_{10}
Press 1/x	0·1	calculates $1/R_{10}$
Press SUM	0·1	adds display to memory
Press RCL	0·3	recalls accumulated memory total
Press 1/x	3·3333333	calculates $1/1/R$. This is the answer.

All of this will seem very tedious at first, but after a little practice will become quick and easy. Try not to press the wrong keys — that is the easiest known method for getting the wrong answer!

Example 1.14

Resistors of 16 Ω, 24 Ω and 48 Ω, respectively, are connected in parallel to a 240 V supply. Calculate the total current.

There are two ways of solving this problem.

Method 1

Find the equivalent resistance and use it with the supply voltage to find the total current.

$$\frac{1}{R} = \frac{1}{16} + \frac{1}{24} + \frac{1}{48} = \frac{3+2+1}{48} = \frac{6}{48}$$

$$R = \frac{48}{6} \text{ ohms} = 8 \text{ } \Omega$$

$$I = \frac{U}{R} = \frac{240}{8} \text{ amperes} = 30 \text{ A}$$

Method 2

Find the current in each resistor. Add these currents to give the total current.

$$\text{current in 16 } \Omega \text{ resistor} = \frac{U}{R_{16}} = \frac{240}{16} \text{ amperes} = 15 \text{ A}$$

$$\text{current in 24 } \Omega \text{ resistor} = \frac{U}{R_{24}} = \frac{240}{24} \text{ amperes} = 10 \text{ A}$$

$$\text{current in 48 } \Omega \text{ resistor} = \frac{U}{R_{48}} = \frac{240}{48} \text{ amperes} = 5 \text{ A}$$

$$\text{total current} = 15 + 10 + 5 \text{ amperes} = 30 \text{ A}$$

There are many applications of parallel circuits. The elements of a two-bar fire are connected in parallel, and the heat output is varied by switching one bar on or off as required. The two circuits in a cooker grill can be connected in three ways to give three-heat control, as indicated in the following example.

Example 1.15

The grill of an electric cooker has two identical elements, each of resistance 48 Ω, which are connected in parallel for 'high' heat and in series for 'low' heat. One element only is used for 'medium' heat. Calculate the current drawn from a 240 V supply for each switch position.

Low   The elements are in series (Figure 1.11*a*)

Total resistance = 48 + 48 ohms = 96 Ω

Therefore $I = \dfrac{U}{R} = \dfrac{240}{96}$ amperes = 2·5 A

Medium   One element only in use (Figure 1.11*b*)

Therefore $I = \dfrac{U}{R} = \dfrac{240}{48}$ amperes = 5 A

High   The elements are in parallel (Figure 1.11*c*)

Total resistance $= \dfrac{48}{2}$ ohms = 24 Ω

Therefore $I = \dfrac{U}{R} = \dfrac{240}{24}$ amperes = 10 A

If only two resistors are connected in parallel, there is a simpler way of calculating the equivalent resistance than by adding the reciprocals. It

Figure 1.11   3-heat switching circuits for Example 1.15
> *a* Low
> *b* Medium
> *c* High

is particularly useful where resistor values are not whole numbers, and when therefore the lowest common denominator is not easily found. It can be shown that, for two resistors in parallel,

$$R = \frac{R_1 \times R_2}{R_1 + R_2}$$

This method is often referred to as 'product over sum'. It should be noticed that it is only useful for TWO resistors. Although there are equivalent formulas for more resistors, they are so complicated as to be difficult to use. However, if there *are* more than two resistors, they can be considered in stages.

For example, consider that we need to find the equivalent resistance of three resistors, of 15 Ω, 10 Ω and 4 Ω all connected in parallel. Take the first two resistors and find their equivalent value, which we will call R_A.

$$R_A = \frac{R_1 \times R_2}{R_1 + R_2}$$

$$= \frac{10 \times 15}{10 + 15} \text{ ohms}$$

$$= \frac{150}{25} \text{ ohms}$$

$$= 6 \ \Omega$$

Now take this equivalent 6 Ω resistor and put it in parallel with the third, 4 Ω, resistor.

$$R = \frac{R_A \times R_3}{R_A + R_3}$$

$$= \frac{6 \times 4}{6 + 4} \text{ ohms}$$

$$= \frac{24}{10} \text{ ohms}$$

$$= 2 \cdot 4 \ \Omega$$

1.12   Series-parallel circuits

A series–parallel circuit is one which is made up of series and parallel parts in combination. The possible number of combinations is endless, but all these circuits can be solved by simplification. A number of resistors, seen to be in series or in parallel, are replaced by one resistor which has the same effect on the circuit. This principle is explained in Example 1.16.

Example 1.16

Two banks of resistors are connected in series. The first bank consists of two resistors of 10 Ω and 40 Ω in parallel, and the second consists of three resistors, each of 12 Ω, connected in parallel. What is the resistance of the combination, and what current will be taken from a 12 V supply to which it is connected?

The circuit diagram is shown in Figure 1.12a. To solve, we must look for groups of resistors connected in series or in parallel. The first bank consists of two resistors in parallel, so we must find the resistance of this combination.

$$\frac{1}{R} = \frac{1}{10} + \frac{1}{40} = \frac{4+1}{40} = \frac{5}{40}$$

Therefore
$$R = \frac{40}{5} \text{ ohms} = 8 \ \Omega$$

Thus, the first group of resistors can be replaced by a single resistor of 8 Ω.

The second bank consists of three 12 Ω resistors in parallel. Its resistance can be found thus:

$$\frac{1}{R} = \frac{1}{12} + \frac{1}{12} + \frac{1}{12} = \frac{3}{12}$$

Therefore
$$R = \frac{12}{3} \text{ ohms} = 4 \ \Omega$$

The second group can thus be replaced by a single 4 Ω resistor. Figure 1.12b shows a simple series circuit which is the equivalent of that in Figure 1.12a. These two circuits are not identical, but they have the same resistance and will take the same current when connected to a supply.

These two resistors can now be combined to a single equivalent resistor.

$$R = 8 + 4 \text{ ohms} = 12 \ \Omega$$

Figure 1.12   Circuit diagrams for Example 1.16

This is the resistance of the complete circuit. To find the current, Ohm's law is applied:

$$I = \frac{U}{R} = \frac{12}{12} = 1 \text{ A}$$

Example 1.17

Figure 1.13 shows a resistor network connected to a 240 V supply. Ammeters and voltmeters are to be connected to measure the current in each resistor, and the PD across each resistor. Redraw the diagram, adding these instruments. Calculate their readings.

Ammeters must be connected so that the current to be measured passes through them. An ammeter must therefore be connected in series with each resistor.

Voltmeters must be connected so that the potential difference to be measured is also across their terminals. A voltmeter could thus be connected in parallel with each resistor. However, resistors connected in parallel clearly have the same potential difference across them, so one voltmeter connected across such a parallel group will suffice.

Figure 1.14a shows the circuit with ammeters and voltmeters added. It should be noted that voltmeter V_X is connected across the 80 Ω resistor *and* ammeter A_1, as well as across the 120 Ω resistor *and* ammeter A_2. In fact, each of the ammeters has resistance and hence a potential difference appears across it when carrying current. Thus the voltmeter will read the sum of the PDs across a resistor and an ammeter. In practice, the resistance of an ammeter is so small that, in most cases, the potential difference across it is small and can be ignored.

To find the readings on the instruments, we must calculate the PD across each resistor, as well as the current through each resistor.

The first step is to find the resistance of the whole circuit, and hence the total current. It is most important when dealing with circuits of this sort to be quite clear which part of the circuit is being considered. To this end, the three series-connected sections of the circuit have been called X, Y and Z, respectively (Figure 1.14a). As in the previous example, the circuit must now be reduced to the simple series circuit shown in Figure 1.14b.

Figure 1.13   Circuit for Example 1.17

Figure 1.14   Circuits for solution of Example 1.17

$$\frac{1}{R_x} = \frac{1}{80} + \frac{1}{120} = \frac{3+2}{240} = \frac{5}{240}$$

Therefore
$$R_x = \frac{240}{5} \text{ ohms} = 48 \ \Omega$$

$$R_y = 17 \ \Omega$$

$$\frac{1}{R_z} = \frac{1}{30} + \frac{1}{45} + \frac{1}{90} = \frac{3+2+1}{90} = \frac{6}{90}$$

Therefore
$$R_z = \frac{90}{6} \text{ ohms} = 15 \ \Omega$$

$$\text{Total circuit resistance} = R_x + R_y + R_z$$

$$= 48 + 17 + 15 \text{ ohms}$$

$$= 80 \ \Omega$$

Current from supply, $I = \dfrac{U}{R} = \dfrac{240}{80}$ amperes $= 3$ A

When this current flows in the circuit of Figure 1.14b, the voltage drop across R_x,

$$U_x = IR_x = 3 \times 48 \text{ volts} = 144 \text{ V}$$

Since voltmeter V_x is connected across a group of resistors having the same effect as R_x, V_x will read 144 V.

Similarly, $$U_y = IR_y = 3 \times 17 = 51 \text{ V}$$

and voltmeter V_y will read 51 V

Similarly, $$U_z = IR_z = 3 \times 15 \text{ volts} = 45 \text{ V}$$

and voltmeter V_z will read 45 V

Note: As pointed out in Section 1.10, for a series circuit, the sum of the PDs across the individual resistors is equal to the supply voltage.

To check this,

$$144 \text{ V} + 51 \text{ V} + 45 \text{ V} = 240 \text{ V}$$

We can now calculate the current in each resistor.

The resistors in section X each have the section PD of 144 V applied to them.

Thus $$I_1 = \frac{U_x}{R_1} = \frac{144}{80} \text{ amperes} = 1 \cdot 8 \text{ A}$$

so, ammeter A_1 reads $1 \cdot 8$ A

$$I_2 = \frac{U_x}{R_2} = \frac{144}{120} \text{ amperes} = 1 \cdot 2 \text{ A}$$

and ammeter A_2 reads $1 \cdot 2$ A

Note: As pointed out in Section 1.11, for a parallel circuit, the sum of the currents in individual resistors is equal to the supply current. To check this,

$$1 \cdot 8 \text{ A} + 1 \cdot 2 \text{ A} = 3 \text{ A}$$

The 17 Ω resistor in section Y has 51 V applied to it.

$$I_3 = \frac{U_y}{R_3} = \frac{51}{17} \text{ amperes} = 3 \text{ A}$$

Thus ammeter A_3 reads 3 A.

Inspection of the circuit would have confirmed this current without calculation. Since the 17 Ω resistor has no other resistor in parallel with it, it must carry the whole of the circuit current.

The resistors in section Z each have 45 V applied to them.

Thus $$I_4 = \frac{U_z}{R_4} = \frac{45}{30} \text{ amperes} = 1 \cdot 5 \text{ A}$$

and ammeter A_4 reads $1 \cdot 5$ A

$$I_5 = \frac{U_z}{R_5} = \frac{45}{45} \text{ amperes} = 1 \text{ A}$$

and ammeter A_5 reads 1 A

$$I_6 = \frac{U_z}{R_6} = \frac{45}{90} \text{ amperes} = 0 \cdot 5 \text{ A}$$

and ammeter A_6 reads $0 \cdot 5$ A

Check: $1 \cdot 5$ A $+ 1 \cdot 0$ A $+ 0 \cdot 5$ A $= 3$ A

Example 1.18

The circuit shown in Figure 1.15 takes a current of 6 A from the 50 V supply. Calculate the value of the resistor R_4.

There are several ways of tackling this problem. One of the simplest demonstrates a different approach from those possible in the previous worked examples.

First, mark the currents on the diagram (this has already been done in Figure 1.15).

R_1 has the 50 V supply directly across it, so

$$I_1 = \frac{U}{R_1} = \frac{50}{25} \text{ amperes} = 2 \text{ A}$$

Since the total current is 6 A, it follows that the current I_2 in the upper part of the circuit must be equal to 6 A minus 2 A, or 4 A. The PD across R_2, $U_2 = I_2R_2 = 4 \times 2 \cdot 5$ V $= 10$ V.

The parallel combination of R_3 and R_4 is in series with R_2 across the 50 V supply. Since the PD across R_2 is 10 V, the PD across both R_3 and R_4 must be 50 V minus 10 V, or 40 V.

Thus $$I_3 = \frac{U_3}{R_3} = \frac{40}{30} \text{ amperes} = 1 \cdot 33 \text{ A}$$

The currents I_3 and I_4 must sum to the current I_2, *so*

$$I_4 = I_2 - I_3 = 4 - 1 \cdot 33 \text{ amperes} = 2 \cdot 67 \text{ A}$$

Figure 1.15 Diagram for Example 1.18

Applying Ohm's law to R_4,

$$R_4 = \frac{U_4}{I_4} = \frac{40}{2 \cdot 67} \text{ ohms} = 15 \ \Omega$$

Note that, in the working above, $1\tfrac{1}{3}$ A would have been exact whereas $1 \cdot 33$ A is correct to only two decimal places. In general, however, we avoid the use of fractions in electrical calculations, because instruments are invariably scaled decimally. Thus an ammeter could perhaps be read as $1 \cdot 33$ A but not as $1\tfrac{1}{3}$ A.

There are many circuits, often called networks, which are not series–parallel connections and to which the methods indicated will not apply. These must be solved by more advanced circuit theorems, which are beyond our scope at this stage.

However, the vast majority of the circuits encountered by the electrical craftsman are of the series, parallel or series–parallel type.

1.13   Summary of formulas for Chapter 1

$$Q = It \qquad I = \frac{Q}{t} \qquad t = \frac{Q}{I}$$

where Q = electric charge, C
  I = current, A
  t = duration of current, s

$$W = dF \qquad d = \frac{W}{F} \qquad F = \frac{W}{d}$$

where W = energy used or work done, mN or J
  F = force applied, N
  d = distance moved, m

$$P = \frac{W}{t} \qquad W = Pt \qquad t = \frac{W}{P}$$

where P = power or rate of doing work, W

$$W = UQ \qquad Q = \frac{W}{U} \qquad U = \frac{W}{Q}$$

where U = potential difference, V

For resistors in series,

$$R = R_1 + R_2 + R_3 \ldots$$

For resistors in parallel,

$$\frac{1}{R} = \frac{1}{R_1} + \frac{1}{R_2} + \frac{1}{R_3} + \ldots$$

where R = total circuit resistance, Ω
  R_1, R_2, R_3 etc. = individual circuit resistances, Ω

$$I = \frac{U}{R} \qquad R = \frac{U}{I} \qquad U = IR$$

where U = applied voltage, V
       I = current, A
       R = circuit resistance, Ω

1.14   Exercises

1 A current of 10 A flows for two minutes. What charge is transferred?

2 For how long must a current of 4 mA flow so as to transfer a charge of 24 C?

3 What current must flow if 100 C is to be transferred in 8 seconds?

4 Briefly describe one example of each of the chemical, heating and electromagnetic effects of an electric current.

5 A force of 450 N is required to lift a bundle of conduit. How much work is done if it is raised from the floor to the roof rack of a van 2 m high?

6 Energy of 2 J is required to close a contactor against a spring exerting a force of 80 N. How far does the contactor move?

7 When an electric motor is pushed 30 m across a level floor, 4500 J of work is done. What force is needed to move the motor?

8 A force of 35 N is required at the end of a spanner 0·2 m long to move a nut on a thread. How much work is done in giving the nut one complete turn?

9 A DC generator has an EMF of 200 V and provides a current of 10 A. How much energy does it provide each minute?

10 A photocell causes a current of 4 μA in its associated circuit, and would take 1000 days to dissipate an energy of 1 mJ. What EMF does it provide?

11 An electric blanket is required to provide heat energy at the rate of 7200 J/min from a 240 V supply. What current will flow?

12 If the total resistance of an earth fault loop is 4 Ω, what current will flow in the event of a phase-to-earth fault from 240 V mains?

13 During a flash test, a voltage of 20 kV is applied to a cable with an insulation resistance of 5 MΩ. What will be the leakage current?

14 What is the resistance of an immersion-heater element that takes 12·0 A from a 240 V supply?

15 An indicator lamp has a hot resistance of 50 Ω and a rated current of 0·2 A. What is its rated voltage?

16 The PD across an earth-continuity conductor of resistance 0·3 Ω is found to be 4·5 V. What current is the conductor carrying?

17 Four resistors of values 5 Ω, 15 Ω, 20 Ω and 40 Ω, respectively, are connected in series to a 240 V supply. Calculate the resulting current and the PD across each resistor.

18 A 6 Ω resistor and a resistor of unknown value are connected in series to a 12 V supply, when the PD across the 6 Ω resistor is measured as 9 V. What is the value of the unknown resistor?

19 Calculate the resistance of the element of a soldering iron that takes $0 \cdot 5$ A from 240 V mains when connected to them by cables having a total resistance of $0 \cdot 2$ Ω.

20 Calculate the equivalent resistance of each of the following parallel-connected resistor banks:
  (*a*) 2 Ω and 6 Ω;
  (*b*) 12 MΩ, 6 MΩ and 36 MΩ; and
  (*c*) 100 μΩ, 600 μΩ and $0 \cdot 0012$ Ω.

21 Answer the following questions by writing down the missing word or words:
  (*a*) A fuse protects a circuit against......and uses the......effect of an electric current.
  (*b*) The unit of quantity of electricity is called the......
  (*c*) Two good conductors of electrical current are......and......
  (*d*) Two items of electrical equipment that use the electromagnetic effect are......and ......
  (*e*) Two insulating materials used in the electrical industry are...... and......
  (*f*) In an electrical circuit, the electron flow is from the......terminal to the......terminal.
  (*g*) A bimetallic strip uses the......effect of an electric current.
  (*h*) The electron has a......charge.
  (*i*) An EMF of 72 V is applied to a circuit, and a current of 12 A flows. The resistance of the circuit is......
  (*j*) The effective resistance of two 10 Ω resistors connected in parallel is......
  (*k*) Quantity of electricity = ... × ......
  (*l*) $0 \cdot 36$ amperes = ...milliamperes.
  (*m*) $3 \cdot 3$ kilovolts = ...volts.

22 Three resistors are connected in parallel across a supply of unknown voltage. Resistor A is of $7 \cdot 5$ Ω and carries a current of 4 A. Resistor B is of 10 Ω, and resistor C is of unknown value but carries a current of 10 A. Calculate the supply voltage, the current in resistor B and the value of resistor C.

23 Three parallel-connected busbars have resistances of $0 \cdot 1$ Ω, $0 \cdot 3$ Ω and $0 \cdot 6$ Ω, respectively, and, in the event of a short-circuit, would be connected directly across a 400 V supply. Calculate the equivalent resistance of the combination, the total fault current and the current in each busbar.

24 Resistors of 7 Ω, 14 Ω and 14 Ω, respectively, are connected in parallel. This bank is connected in series with a $2 \cdot 5$ Ω resistor across a supply of unknown PD. The current flow in the $2 \cdot 5$ Ω resistor is 2 A. What is the supply voltage?

25 (*a*) Show, by separate drawings for 'high', 'medium' and 'low' heat positions, the connections of a series–parallel switch controlling two separate sections of resistance wire forming the element of a heating appliance.
  (*b*) If the two sections of resistance wire are of equal resistance, what is the

proportional current flow and heating effect in the 'medium' and 'low' positions relative to the 'high' position?

26 Three resistors, having resistances of $4 \cdot 8 \ \Omega$, $8 \ \Omega$ and $12 \ \Omega$, all connected in parallel, are supplied from a 48 V supply. Calculate the current through each resistor and the current taken from the supply. Calculate the effective resistance of the group.

27 Two resistances of $4 \ \Omega$ and $12 \ \Omega$ are connected in parallel with each other. A further resistance of $10 \ \Omega$ is connected in series with the combination. Calculate the respective direct voltages which should be applied across the whole circuit,
    (*a*) to pass 6 A through the $10 \ \Omega$ resistance
    (*b*) to pass 6 A through the $12 \ \Omega$ resistance.

28 (*a*) State Ohm's law in words and by symbols.
    (*b*) Three relay coils, each of $7 \cdot 5 \ \Omega$ resistance, are connected in parallel. What current would flow in the circuit when the supply is 12 V?
    (*c*) Calculate the resistance required to limit the current in a circuit to $4 \cdot 8$ A with a voltage of 200 V.

29 Three banks of resistors are connected in series across a 240 V supply. Bank A consists of three resistors R_1, R_2 and R_3, each of resistance $60 \ \Omega$, connected in parallel. Bank B comprises two resistors, R_4 of resistance $40 \ \Omega$ and R_5 of resistance $120 \ \Omega$, connected in parallel. Bank C has three parallel-connected resistors, $R_6 = 50 \ \Omega$, $R_7 = 100 \ \Omega$ and $R_8 = 300 \ \Omega$. Calculate
    (*a*) the resistance of the complete circuit
    (*b*) the current in each resistor
    (*c*) the PD across each resistor.

30 A resistance network is connected as shown in Figure 1.16, and takes a total current of $2 \cdot 4$ A from the 24 V supply. Calculate the value of the resistor R_X.

31 The voltmeter shown in Figure 1.17 reads 39 V. What is the value of the resistor across which the voltmeter is connected?

32 Three banks of resistors are connected in series. Bank 1 comprises $40 \ \Omega$ and $24 \ \Omega$ in parallel; bank 2 comprises a single resistor of $5 \ \Omega$; bank 3 comprises four resistors in parallel, each of $16 \ \Omega$.

    Ammeters are inserted to measure the current through each resistor, and voltmeters to measure the voltage across each bank. Sketch a layout of the apparatus, and calculate the reading on each instrument if 48 V be applied to the circuit.

33 A bank of three paralleled resistors each of $12 \ \Omega$ is connected in series with another bank consisting of three resistors in parallel, their values being $4 \ \Omega$, $4 \ \Omega$ and $2 \ \Omega$. If a 100 V supply is applied across this circuit, what current would flow through the $2 \ \Omega$ resistor?

1.15   Multiple-choice exercises

1M1 The central part of an atom is called the
    (*a*) neutron   (*b*) nucleus   (*c*) proton   (*d*) electron

Figure 1.16 Circuit diagram for Exercise 1.30

Figure 1.17 Circuit diagram for Exercise 1.31

1M2 An electric current is considered to consist of
(a) a drift of electrons from negative to positive
(b) a movement of positive ions from positive to negative
(c) an electric fluid travelling from positive to negative
(d) a drift of electrons from positive to negative.

1M3 The unit for the quantity of electricity is called the
(a) ampere   (b) watt   (c) joule   (d) coulomb

1M4 The heating effect of an electrical current can be used in
(a) the fuel cell   (b) the transformer
(c) an ammeter   (d) an electric kettle

1M5 The element nickel is
(a) a soft metal   (b) an electrical conductor
(c) a kind of money   (d) an electrical insulator

1M6 The symbol for energy and work is
(a) J   (b) P   (c) U   (d) W

1M7 The symbol U is used to denote
(a) electromotive force or potential difference
(b) a material unsuitable for electrical use
(c) electrical resistivity
(d) a voltmeter

1M8 The formula which relates resistance, potential difference and current is

(a) $I = UR$ (b) $R = \dfrac{I}{U}$ (c) $R = \dfrac{U}{I}$ (d) $R_T = R_1 + R_2$

1M9 If a current of 26 mA flows in a circuit when a 24 V supply is connected to it, its resistance is
(*a*) 923 Ω (*b*) 108 Ω (*c*) 9·23 kΩ (*d*) 92·3 Ω

1M10 To read a current correctly an ammeter must be connected
(*a*) so that no current passes through it
(*b*) directly across the supply
(*c*) in parallel with the circuit concerned
(*d*) in series with the circuit concerned

1M11 Two resistors connected in series have a combined resistance of 47 kΩ. If one has a value of 15 kΩ, the resistance of the other is
(*a*) 62 kΩ (*b*) 32 kΩ (*c*) 22 kΩ (*d*) 3·13 kΩ

1M12 The formula to find the combined resistance of two resistors R_1 and R_2 connected in parallel is

(*a*) $R_T = \dfrac{R_1 \times R_2}{R_1 + R_2}$ (*b*) $R_T = R_1 + R_2$

(*c*) $R_T = \dfrac{R_1 + R_2}{R_1 \times R_2}$ (*d*) $R_T = \dfrac{R_1}{R_2}$

1M13 The combined resistance of 56 Ω, 12 Ω and 33 Ω resistors connected in parallel is
(*a*) 101 Ω (*b*) 7·60 Ω (*c*) 7·07 Ω (*d*) 20·4 Ω

1M14 The resistor arrangement shown in Figure 1.18 is connected
(*a*) in parallel   (*b*) as a network
(*c*) in series   (*d*) in series–parallel

Figure 1.18   Circuit diagram for Exercise 1M14

Figure 1.19   Circuit diagram for Exercises 1M15 and 1M16

1M15 The total resistance of the circuit shown in Figure 1.19 is
    (*a*) 269 Ω   (*b*) 100 Ω   (*c*) 164 Ω   (*d*) 97 Ω

1M16 The current flowing in the 50 Ω resistor in the circuit of Figure 1.19 is
    (*a*) 2·4 A   (*b*) 1·12 A   (*c*) 1·28 A   (*d*) 0·892 A

Chapter 2
Resistance and resistors

2.1 Introduction

The conductors used to connect together the components of a circuit have the property of **resistance**. Usually, this resistance will be so small, in comparison with that of the components themselves, that it may be neglected for the purposes of circuit calculations. The conductor resistance will be present, however, and, sometimes, as indicated in Section 2.7, the volt drop in resistive conductors must be taken into account.

Figure 2.1   Fixed resistors
  a Carbon low-power resistor
  b Wire-wound encapsulated resistor

34

c

d

Figure 2.1   Fixed resistors (cont.)
　　　　c Vitreous enamelled wire-wound resistor
　　　　d High-power asbestos woven-mat resistor

If conductors without any resistance at all were possible at a reasonable cost, they would be used widely.

Probably the most common circuit component is the **resistor**. Such a component is deliberately made to offer resistance to electric current. Resistors can exist in a very wide variety of shapes and sizes, a few of the most common constructions being shown in Figure 2.1. The British Standard objective symbol for a **fixed resistor** is shown in Figure 2.2*a*, with the alternative symbol in Figure 2.2*b*. This symbol has been in use since 1966. The alternative symbol is still in use, and is likely to be so for many years.

In many practical circuits, as well as in laboratory work, it is often necessary to use a **variable resistor**. This is usually a resistor with a sliding contact, so that the resistance between one end and the slider can be adjusted by movement of the latter. Some examples of variable resistors are shown in Figure 2.3. Figure 2.4 shows the objective and alternative general symbols for a variable resistor, and Figure 2.5 shows the symbols that are used when the connection to the slider must be shown. A variable resistor is sometimes called a **rheostat**.

a   b

Figure 2.2   Circuit symbols for fixed resistor
  a Objective symbol   *b* Alternative symbol

Figure 2.3 Variable resistors
    a Wire-wound resistor with adjustable tapping
    b Wire-wound resistor with continuous variation
    c Wire-wound toroidal potential divider

Figure 2.4 Circuit symbols for variable resistor
    a Objective symbol   *b* Alternative symbol

Figure 2.5 Circuit symbols for resistor with moving contact
    a Objective symbol   *b* Alternative symbol

Many variable resistors can be connected as voltage dividers or **potential dividers** (Figure 2.6). The voltage U_1 that appears at the output terminals will depend on the setting of the slider, and on the current taken from the output. Provided that the output current is very small, it follows that

$$U_1 = U \times \frac{R_1}{R}$$

This method of connection is often used to provide a variable-voltage supply.

Figure 2.6   Variable voltage output from potential divider

2.2   Effect of dimensions on resistance

Let us imagine a cube of conducting material (Figure 2.7a) that has a resistance of r ohms between opposite faces. If, say, five of these cubes are joined (Figure 2.7b), they form a series resistor chain which has a total resistance of $5r$ ohms. If, say, four conductors, each made up of five cubes, are placed side by side (Figure 2.7c), the result will be one conductor having four times the cross-sectional area of one of the original conductors. The total resistance R of this composite conductor will be that of four $5r$ ohm resistors connected in parallel:

$$\frac{1}{R} = \frac{1}{5r} + \frac{1}{5r} + \frac{1}{5r} + \frac{1}{5r} = \frac{4}{5r}$$

Therefore
$$R = \frac{5r}{4} \text{ ohms}$$

This is the resistance of a conductor five cubes long and of four cubes in cross-section.

If the conductor had a length l and a uniform cross-sectional area a, its resistance would be

$$R = \frac{r \times l}{a}$$

In this way, the resistance of a conductor can be expressed in terms of its length, its cross-sectional area and the resistance between opposite faces of a cube of material identical to that of which the conductor is made. Conductors are not often shaped so that they can be considered directly as a number of cubes, but the result obtained is true for a cable of any shape if its cross-sectional area is uniform throughout its length.

Example 2.1

The resistance of 50 m of a certain cable is $0 \cdot 15 \; \Omega$. Calculate the resistance of 800 m of this cable.

Figure 2.7   Dependence of conductor resistance on dimensions

From the information above, $R \propto \dfrac{l}{a}$.

For a cable of a given size, the cross-sectional area a is constant, so resistance is directly proportional to length, or $R \propto l$. 800 m of cable is 800/50, or 16 times, longer than 50 m of cable. Thus, the resistance of 800 m of cable will be 16 times that of 50 m of cable.

Therefore $\qquad\qquad R = 16 \times 0 \cdot 15$ ohms $= 2 \cdot 4\Omega$

Example 2.2

If the resistance per 100 m of a cable of cross-sectional area $2 \cdot 5$ mm^2 is $0 \cdot 06\Omega$, what is the resistance of 100 m of a cable made of the same conductor material which has a cross-sectional area of 6 mm^2?

Since l is constant, $R \propto 1/a$

6 mm^2 is $6/2 \cdot 5$ or $2 \cdot 4$ times larger than $2 \cdot 5$ mm^2.

Since resistance is *inversely* proportional to cross-sectional area, the resistance of the second cable will be $2 \cdot 4$ times *smaller* than that of the first cable, or

$$R = \frac{0 \cdot 06}{2 \cdot 4} \text{ ohms} = 0 \cdot 025 \text{ } \Omega$$

Example 2.3

If the resistance of 100 m of a 4 mm^2 cable is $0 \cdot 08$ Ω, what will be the resistance of 700 m of 10 mm^2 cable made of the same conductor material?

$$R \propto \frac{l}{a}$$

Therefore
$$R_2 = R_1 \times \frac{l_2}{l_1} \times \frac{a_1}{a_2}$$

where subscripts 1 and 2 refer to the first and second cables, respectively.

Therefore
$$R_2 = 0 \cdot 08 \times \frac{700}{100} \times \frac{4}{10} \text{ ohms}$$

$$= 0 \cdot 224 \text{ } \Omega$$

Example 2.4

In the manufacture of aluminium cable, a 10 m length of thick circular rod, which has a resistance of $0 \cdot 015 \Omega$, is drawn out until its new diameter is one-fifth of its previous measurement. Calculate the length of the drawn cable and its resistance.

By drawing the aluminium bar through a die, its length is increased by reducing its diameter. The volume must, however, be the same before and after the drawing operation, since no metal is lost. If $V =$ volume,

$$V_1 = l_1 \times a_1 = l_1 \times \frac{\pi d_1^2}{4}$$

and
$$V_2 = l_2 \times a_2 = l_2 \times \frac{\pi d_2^2}{4}$$

But
$$V_1 = V_2, \text{ so } l_1 \times \frac{\pi d_1^2}{4} = l_2 \times \frac{\pi d_2^2}{4}$$

Therefore
$$l_2 = l_1 \times \frac{\pi d_1^2}{4} \times \frac{4}{\pi d_2^2} = l_1 \times \frac{d_1^2}{d_2^2}$$

Since the wire is one-fifth of the initial diameter after drawing,

$$\frac{d_1}{d_2} = \frac{1}{1/5} = 5$$

Therefore $\quad l_2 = 10 \times \dfrac{5^2}{1^2} \text{ metres} = 10 \times 25 \text{ metres} = 250 \text{ m}$

$$R \propto \frac{l}{a} \quad \text{so} \quad R_2 = R_1 \times \frac{l_2}{l_1} \times \frac{a_1}{a_2}$$

$$R_1 = 0 \cdot 015 \ \Omega \qquad l_1 = 10 \text{ m} \qquad l_2 = 250 \text{ m}$$

$$a_1 = \frac{\pi d_1^2}{4} \propto \frac{\pi 1^2}{4} \propto \frac{\pi}{4}$$

$$a_2 = \frac{\pi d_2^2}{4} \propto \frac{\pi}{4} \times \left(\frac{1}{5}\right)^2 \propto \frac{\pi}{4 \times 25}$$

Therefore $\quad R_2 = 0 \cdot 015 \times \dfrac{250}{10} \times \dfrac{\pi}{4} \times \dfrac{4 \times 25}{\pi} \text{ ohms}$

$$= 0 \cdot 015 \times 25 \times 25 \text{ ohms}$$

$$= 9 \cdot 375 \ \Omega$$

2.3 Resistivity

The resistance between opposite faces of a cube of the conductor material, given the value r ohms in Section 2.2, is called the **resistivity**, or, sometimes, the **specific resistance**, of the material. For most conductors, this value is very low, and is usually measured in microhms ($\mu\Omega$) for a cube of given side length. For instance, the resistivity of copper is about $0 \cdot 0172 \ \mu\Omega$ for a 1 metre cube, and is therefore expressed as $0 \cdot 0172 \ \mu\Omega$m. The symbol for resistivity is the Greek letter ρ (pronounced 'rho'), which can now take the place of r in our previous formula, so that

$$R = \frac{\rho l}{a}$$

where $\quad R$ = conductor resistance
$\quad\quad\quad \rho$ = cable resistivity
$\quad\quad\quad l$ = cable length
$\quad\quad\quad a$ = cable cross-sectional area

For this formula to be correctly applied, it should be noticed that

(*a*) the resistance calculated will be in the same units as those in which resistivity is given (usually microhms) (*b*) the units of length and cross-sectional area must match the unit included in the given resistivity; for instance, if resistivity is quoted in microhm metres, length must be in metres and cross-sectional area in square metres.

Table 2.1 gives the resistivities, in three commonly used units, of some common metallic conductors. These are average values, since a slight variation is found to occur depending on conductor condition.

The figures given in Table 2.1 will usually increase as the material temperature increases. This effect will be considered in Section 2.5.

The differing values of resistivity for different conductor materials make it clear that the resistance of a wire or cable will depend on the material of which it is made, as well as on its length and cross-sectional area. Table 2.1 thus shows clearly why copper is so widely used as a conductor material, being second only to silver in its resistivity value. Aluminium is about half as resistive again as copper, but its lightness and cheapness have led to its increasing use for cables and busbars.

Table 2.1: Resistivity of common conductors

Material	Resistivity		
	$\mu\Omega$m	$\mu\Omega$cm	$\mu\Omega$mm
Copper (annealed)	0·0172	1·72	17·2
Copper (hard drawn)	0·0178	1·78	17·8
Aluminium	0·0285	2·85	28·5
Tin	0·114	11·4	114
Lead	0·219	21·9	219
Mercury	0·958	95·8	958
Iron	0·100	10·0	100
Silver	0·0163	1·63	16·3
Brass	0·06–0·09	6–9	60–90

2.4   Resistance calculations

All resistance calculations are based on the formula $R = \rho l/a$, given in Section 2.3. Thus, if resistivity, length and cross-sectional area are known, the resistance may be calculated. In fact, if any three of the four values are known, the fourth may be found.

Example 2.5

Calculate the resistance of 1000 m of 16 mm^2 single annealed copper conductor. The cross-sectional area of the conductor is 16 mm^2.

The three quantities ρ, l and a must be expressed in terms of the same unit of length. If the millimetre is chosen,

$\rho = 17 \cdot 2 \ \mu\Omega\text{mm}$ from Table 2.1, or $\dfrac{17 \cdot 2}{10^6} \ \Omega\text{mm}$

$l = 1000 \ \text{m} = 1000 \times 1000 \ \text{millimetres} = 1 \ 000 \ 000 \ \text{mm} \ \ \text{or} \ \ 10^6 \ \text{mm}$

$a = 16 \ \text{mm}^2$

$$R = \frac{\rho l}{a}$$

$$= \frac{17 \cdot 2}{10^6} \times \frac{10^6}{16} \ \text{ohms}$$

$$= \frac{17 \cdot 2}{16} \ \text{ohms}$$

$$= 1 \cdot 075 \ \Omega$$

As an alternative, we could choose the metre as the standard of length.

$$\rho = 0 \cdot 0172 \ \mu\Omega\text{m} \ \text{from Table 2.1}$$

$$l = 1000 \ \text{m}$$

$$a = 16 \ \text{mm}^2 = \frac{16}{10^6} \ \text{square metres}$$

$$R = \frac{\rho l}{a}$$

$$= \frac{0 \cdot 0172}{10^6} \times 1000 \times \frac{10^6}{16} \ \text{ohms}$$

$$= \frac{17 \cdot 2}{16} \ \text{ohms}$$

$$= 1 \cdot 075 \ \Omega$$

Example 2.6

Calculate the resistance of 1000 m of hard-drawn copper conductor, if ρ for the material is $0 \cdot 0178 \ \mu\Omega\text{m}$ and the cross-sectional area of the conductor is $4 \cdot 7 \ \text{mm}^2$.

Since ρ is given in $\mu\Omega\text{m}$, l must be in metres and a in square metres.

$$R = \frac{\rho l}{a} = \frac{0 \cdot 0178 \times 1000}{10^6} \times \frac{10^6}{4 \cdot 7} \ \text{ohms}$$

$$= \frac{17 \cdot 8}{4 \cdot 7} \ \text{ohms}$$

$$= 3 \cdot 79 \ \Omega$$

Example 2.7

Calculate in metres the length of an aluminium conductor (of cross-sectional area 40 mm^2) which will have a resistance of 0·57 Ω.

From Table 2.1, ρ for aluminium is 0·0285 $\mu\Omega$m, so a must be expressed in m^2, when l will be found in metres.

$$R = \frac{\rho l}{a} \text{ so } l = \frac{Ra}{\rho}$$

Therefore

$$l = 0 \cdot 57 \times \frac{40}{10^6} \times \frac{10^6}{0 \cdot 0285} \text{ metres}$$

$$= \frac{0 \cdot 57 \times 40}{0 \cdot 0285} \text{ metres}$$

$$= 800 \text{ m}$$

Example 2.8

The heating spiral of an electric fire is removed from its silicon tube, uncoiled and measured. It has a length of 4 m, a diameter of 0·2 mm, and its resistance is measured as 20Ω. What is the resistivity, in $\mu\Omega$mm, of the spiral material?

$$R = \frac{\rho l}{a} \text{ so } \rho = \frac{Ra}{l}$$

Since ρ is required in $\mu\Omega$mm, R must be expressed in microhms, a in square millimetres and l in millimetres.

$$a = \frac{\pi d^2}{4}$$

$$= \frac{3 \cdot 142 \times 0 \cdot 2 \times 0 \cdot 2}{4} \text{ square millimetres}$$

$$= 0 \cdot 03142 \text{ mm}^2$$

$$\rho = \frac{Ra}{l}$$

$$= \frac{20 \times 1\,000\,000 \times 0 \cdot 03142}{4000} \text{ microhm millimetres}$$

$$= \frac{628 \cdot 4}{4} \text{ microhm millimetres}$$

$$= 157 \cdot 1 \text{ } \mu\Omega\text{mm}$$

The examples above indicate the steps to be followed in this type of problem, which has many practical applications. The golden rule is to be careful to express resistivity, length and cross-sectional area in terms of the same unit of length.

2.5   Effect of temperature on resistance

The following test will show in a very convincing manner that change of temperature can affect the resistance of a conductor. When connected to a 2 V supply, a 60 W, 240 V lamp takes a current of 25 mA. Its resistance can thus be calculated:

$$R = \frac{U}{I} = \frac{2}{0 \cdot 025} \text{ ohms} = 80 \text{ } \Omega$$

If the lamp is now connected to a 240 V supply, its filament becomes white hot and glows brightly, and will be found to take a current of 250 mA. Its new resistance will thus be

$$R = \frac{U}{I} = \frac{240}{0 \cdot 025} \text{ ohms} = 9600 \text{ } \Omega$$

We can see that the increase in temperature has increased the filament resistance by 12 times! This extreme example is striking proof that temperature affects resistance.

It is often important to be able to calculate the resistance of a conductor at any given temperature. To be able to achieve this simply, we use the **temperature coefficient of resistance** for the material concerned, which is given the symbol α (Greek 'alpha'). The temperature coefficient of resistance of a material at 0°C can be defined as the change in resistance of a 1 Ω sample of the given material, when its temperature is increased from 0°C to 1°C. For instance, an annealed copper conductor, which has a resistance of 1 Ω at 0°C, will have a resistance of 1·0043 Ω at 1°C. Hence we can say that the temperature coefficient of resistance of annealed copper at 0°C is 0·0043 ohm per ohm per degree Celsius, or 0·0043/°C.

In practice, it is often difficult to measure resistances at 0°C, so temperature coefficients of resistance are often expressed for other temperatures, such as 20°C. Table 2.2 gives some temperature coefficients of resistance of some common conductors.

Table 2.2: Temperature coefficients of resistance of some conductors

Material	(/°C at 0°C)	(/°C at 20°C)
Annealed copper	+ 0·0043	+ 0·00396
Hard-drawn copper	+ 0·0043	+ 0·00396
Aluminium	+ 0·0040	+ 0·00370
Brass	+ 0·0010	+ 0·00098
Iron	+ 0·0066	+ 0·00583
Nickel–chromium	+ 0·00017	+ 0·000169
Carbon (graphite)	− 0·0005	− 0·00047
Silver	+ 0·0041	+ 0·00379

Carbon has a negative temperature coefficient of resistance, which means that its resistance decreases as temperature increases.

Provided that the temperature coefficient of resistance of a conductor material is known, a simple formula can be deduced for calculating the conductor resistance at any temperature from the resistance at $0°C$. Let R_0 be the resistance of a conductor at $0°C$, and let α be its temperature coefficient of resistance in '/°C' at $0°C$.

From the definition of temperature coefficient of resistance, the change in resistance will be R_0 ohms at $1°C$, $2R_0$ ohms at $2°C$ and tR_0 ohms at t degree Celcius. If R_t is the total conductor resistance at t,

$$R_t = R_0 + R_0\alpha t$$

or

$$R_t = R_0(1 + \alpha t)$$

If the temperature coefficient of resistance at, say, $20°C$ were being used, the formula could be written as $R_t = R_{20}(1 + \alpha t)$, where R_{20} is the conductor resistance at $20°C$, t is the change of temperature from $20°C$, and α is the temperature coefficient of resistance at $20°C$.

Example 2.9

The resistance of 1000 m of $2\cdot5$ mm^2 annealed copper conductor is $5\cdot375\ \Omega$ at $20°C$. Find its resistance at $50°C$.

From Table 2.2, α is $0\cdot00396/°C$ at $20°C$.

Therefore $\qquad R_t = R_0(1 + \alpha t)$

$$= 5\cdot375\{1 + 0\cdot00396\ (50-20)\}\ \text{ohms}$$

$$= 5\cdot375 \times 1\cdot1188\ \text{ohms} = 6\ \Omega$$

One method of temperature measurement is to subject a resistor to the unknown temperature and accurately measure its resistance, hence calculating the temperature from a knowledge of the temperature coefficient of resistance of the resistor material.

Example 2.10

A resistance thermometer with a temperature coefficient of $0\cdot001/°C$ at $0°C$ and a resistance of $3\ \Omega$ at $0°C$ is placed in the exhaust gases of an oil-fired furnace. The thermometer resistance rises, reaching a final steady value of $5\cdot25\ \Omega$. What is the exhaust-gas temperature?

$$R_t = R_0(1 + \alpha t)$$

$$5\cdot25 = 3(1 + 0\cdot001t)$$

$$5\cdot25 = 3 + 0\cdot003t$$

$$5\cdot25 - 3 = 0\cdot003t$$

Therefore
$$t = \frac{2 \cdot 25}{0 \cdot 003} \ °C = 750 °C$$

There are many practical cases where the temperature coefficient of resistance of a material is known at a given temperature, but where it is impracticable to measure the actual resistance of some particular conductor at this temperature. If R_1 is the conductor resistance at temperature t_1, and R_2 the resistance at temperature t_2,

$$R_1 = R_0(1 + \alpha t_1) \quad \text{and} \quad R_2 = R_0(1 + \alpha t_2)$$

Dividing these equations, we have

$$\frac{R_1}{R_2} = \frac{R_0(1 + \alpha t_1)}{R_0(1 + \alpha t_2)} = \frac{1 + \alpha t_1}{1 + \alpha t_2}$$

Hence
$$R_2 = \frac{R_1(1 + \alpha t_2)}{1 + \alpha t_1}$$

and R_0, the resistance at $0°C$, is not needed.

Example 2.11

The field winding of a DC motor is of annealed copper, and has a resistance of 500 Ω at $15°C$. What field current will flow at the operating temperature of $35°C$ if the field PD is 300 V?

From Table 2.2, α for annealed copper is $0 \cdot 0043/°C$ at $0°C$.

$$R_2 = \frac{R_1(1 + \alpha t_2)}{1 + \alpha t_1}$$

$$= \frac{500(1 + 0 \cdot 0043 \times 35)}{1 + 0 \cdot 0043 \times 15} \ \text{ohms}$$

$$= \frac{500 \times 1 \cdot 1505}{1 \cdot 0645} \ \text{ohms}$$

$$= 540 \ \Omega$$

$$I = \frac{U}{R} = \frac{300}{540} \ \text{ampere} = 0 \cdot 556 \ \text{A}$$

2.6   Effects of temperature changes

There are some applications, such as the resistance thermometer considered in Example 2.10, where a resistive conductor is deliberately subjected to temperature changes. Resistors of special materials which give a considerable change in resistance for a small temperature change are often used, and are called **thermistors**. These devices are used increasingly in temperature-measuring equipment, and in conjunction with electronic circuits in special types of thermostat.

The change in resistance of a filament lamp described at the beginning of Section 2.5 illustrates clearly how the change in resistance may cause problems. Taking the 60 W, 240 V lamp mentioned, the cold resistance is 80 Ω. At the instant of switching on the 240 V supply, the current to the lamp will be given by

$$I = \frac{U}{R} = \frac{240}{80} \text{ amperes} = 3 \text{ A}$$

At this instant, the lamp will be taking about 12 times its normal current! The lamp quickly heats up, increasing in resistance as it does so and reducing the current to its normal value of 250 mA. The short-lived heavy current, called a **transient**, can cause difficulties. The magnetic forces set up within the lamp itself may cause it to fail at the instant of switching, and a weak fuse may blow when a bank of lamps served by it is switched on.

This effect is not often noticeable because of the comparatively small currents taken by filament lamps. It would be serious, however, if the element of an electric fire were wound with tungsten, which is the material used for lamp filaments. The normal running current of a 1 kW fire element connected to a 240 V supply is about $4 \cdot 2$ A, but would be about 40 A at the instant of switching on a tungsten element. A fire element takes some time to reach its final temperature, so the heavy current would pass through the circuit for a period long enough to blow the circuit fuse.

To prevent this state of affairs, the elements of heavily loaded heating equipment, such as fires, cookers and immersion heaters, are wound with a special alloy of nickel and chromium which has a near-zero temperature coefficient of resistance. This means that there will be almost no change in the resistance of the element as its temperature increases, and hence virtually no change in the circuit current.

There are many circumstances in which changes in the temperature of electrical equipment will affect its operation. For example, the windings of an electric motor will increase in temperature when it is used. This will result in an increase in the resistance of the winding, and a reduction in the current carried, which may affect the operation of the motor. For instance, reduction of current in the field winding of a direct-current motor will result in an increase in its speed.

The resistance of cables will increase with increasing temperature, giving an increased cable-voltage drop (Section 2.7). Usually, however, the cable resistance will form a very small proportion of total circuit resistance, and the effect will be unimportant. The increase in cable resistance when carrying fault current may have a significant effect on fault loop impedance and on the safety of people using an electrical installation.

2.7   Voltage drop in cables

When choosing cables for an installation, it is necessary to ensure that they will carry the load current without overheating. An equally important factor which must be considered is the voltage drop that will occur in them owing to current flowing through their resistance. If this voltage drop is excessive, the potential difference across the load will be low and efficient operation may be prevented. 'Requirements for Electrical Installations' (*BS 7671*), published by the Institution of Electrical Engineers, lay down the allowable voltage drop from the supply position to any point in the installation.

Since any voltage can be expressed by multiplying a current by a resistance ($U = IR$), voltage drop in a cable depends on the current it carries and on its resistance, the latter depending on conductor material, cable length, and cross-sectional area as well as its temperature (see Section 2.3). The resistance of a given length of cable, and hence its voltage drop for a given current, can be reduced by replacing it with a cable having a larger cross-sectional area.

Example 2.12

A twin $2 \cdot 5$ mm^2 MI cable feeds a heater which takes a current of 20 A. If the cable is 100 m long, calculate the voltage drop in it, and the PD across the heater if the supply voltage is 240 V. What must be the minimum cross-sectional area (CSA) of a replacement cable if the voltage drop is not to exceed 6 V?

MI cables have hard-drawn copper cores, so ρ can be taken from Table 2.1 as $17 \cdot 8$ µΩmm. The cross-sectional area of each conductor is $2 \cdot 5$ mm^2, and the total conductor length will be 200 m.

Cable resistance $\qquad R = \dfrac{\rho l}{a}$

Therefore $\qquad R = \dfrac{17 \cdot 8 \times 200 \times 1000}{2 \cdot 5}$ microhms

$\qquad\qquad\qquad R = 1\ 424\ 000\ \mu\Omega$ or $R = 1 \cdot 424\ \Omega$

Cable voltage drop $\quad U = I \times R$

$\qquad\qquad\qquad\qquad = 20 \times 1 \cdot 424$ volts

$\qquad\qquad\qquad\qquad = 28 \cdot 5$ V

PD across heater $\qquad 240 - 28 \cdot 5$ volts $= 211 \cdot 5$ V

Clearly, the heater, designed for operation at 240 V, would be very inefficient with such a low applied voltage. For this reason, *BS 7671* limits the maximum allowable voltage drop in a circuit to 4% of the supply voltage, or $9 \cdot 6$ V for a 240 V supply.

A voltage drop of $28 \cdot 5$ V is far in excess of the allowable limit of $9 \cdot 6$ V, even though the cable current-carrying capacity of 38 A is not exceeded. If the voltage drop must not exceed $9 \cdot 6$ V, the maximum cable resistance can be found thus:

$$R = \frac{U}{I} = \frac{9 \cdot 6}{20} \text{ ohms} = 0 \cdot 48 \; \Omega$$

$$R = \frac{\rho l}{a} \quad \text{so} \quad a = \frac{\rho l}{R}$$

Minimum CSA $\qquad a = \dfrac{17 \cdot 8 \times 200 \times 1000}{1\,000\,000 \times 0 \cdot 48} \text{ square millimetres}$

$$= 7 \cdot 42 \text{ mm}^2$$

This is the minimum CSA to satisfy the voltage-drop requirements of the Regulations, but it is not a standard size. The next standard size above this is 10 mm^2 twin MI cable (from *BS 7671* tables). The fact that the current rating of this cable is 77 A, whereas it is only required to carry 20 A, illustrates the importance of considering voltage drop as well as current-carrying capacity when choosing cables.

Tables of cable current ratings in *BS 7671* include figures for the voltage drop per ampere per metre length of run for the cable concerned under the conditions indicated. The BS tables are more accurate than the method in the above example, since, for AC circuits, they take into account cable reactance as well as resistance. Reactance and its combination with resistance will be considered in Chapter 10.

In some circumstances, it may be necessary to calculate the resistance of a supply cable to find the voltage drop in it. This process is illustrated in the following examples.

Example 2.13

A motor takes 45 A from a 240 V supply. The motor is fed from the supply by a twin aluminium cable 40 m long, each core of the cable having a cross-sectional area of 25 mm^2. Calculate the voltage at the motor terminals.

The first step is to find the cable resistance. The total length of conductor ('go' and 'return') is 2×40 m $= 80$ m, or 80 000 mm, and, from Table 2.1, the resistivity of aluminium is $28 \cdot 5 \; \mu\Omega$mm.

$$R = \frac{\rho l}{a}$$

$$= \frac{28 \cdot 5 \times 80\,000}{1\,000\,000 \times 25} \text{ ohms}$$

$$= 0 \cdot 0912 \; \Omega$$

$$\text{voltage drop} = I \times R$$

$$= 45 \times 0 \cdot 0912 \text{ volts}$$

$$= 4 \cdot 1 \text{ V approximately}$$

$$\text{motor voltage} = \text{supply voltage} - \text{voltage drop}$$

$$= 240 - 4 \cdot 1 \text{ volts}$$

$$= 235 \cdot 9 \text{ V}$$

Example 2.14

An industrial heater is fed from the supply by a twin annealed-copper cable with $2 \cdot 5$ mm^2 conductors. The current to the heater is 20 A, and the supply voltage is 240 V. If the potential difference at the heater terminals is $235 \cdot 5$ V, calculate the length of the supply cable to the nearest metre.

Here, we must first find the cable voltage drop, and thus its resistance.

$$\text{voltage drop} = 240 - 235 \cdot 5 \text{ volts}$$

$$= 4 \cdot 5 \text{ V}$$

$$\text{cable resistance} = \frac{\text{voltage drop}}{\text{current}}$$

$$= \frac{4 \cdot 5}{20} \text{ ohms}$$

$$= 0 \cdot 225 \ \Omega$$

$$R = \frac{\rho l}{a}, \text{ so, changing the subject of the formula,}$$

$$l = \frac{Ra}{\rho}$$

Therefore

$$l = \frac{0 \cdot 225 \times 2 \cdot 5 \times 10^6}{17 \cdot 2} \text{ millimetres}$$

(from Table 2.1, ρ for annealed copper is $17 \cdot 2$ $\mu\Omega$mm)

$$l = \frac{0 \cdot 225 \times 2 \cdot 5}{17 \cdot 2} \times 10^3 \text{ metres}$$

$$= 32 \cdot 7 \text{ m}$$

This is the total conductor length for the twin cable. The cable length is half the conductor length, which, to the nearest metre, is 16 m.

2.8 Summary of formulas for Chapter 2

$$R = \frac{\rho l}{a} \qquad a = \frac{\rho l}{R} \qquad \rho = \frac{Ra}{l} \qquad l = \frac{Ra}{\rho}$$

where R = conductor resistance, $\mu\Omega$
ρ = conductor resistivity, $\mu\Omega$m or $\mu\Omega$mm

l = conductor length, m or mm

a = conductor cross-sectional area, m^2 or mm^2

Note that, if, for example, ρ is used in $\mu\Omega$m, l must be expressed in m and a in m^2.

$$R_t = R_0(1 + \alpha t) \qquad R_0 = \frac{R_t}{1 + \alpha t}$$

$$\alpha = \frac{R_t - R_0}{R_0 t} \qquad t = \frac{R_t - R_0}{R_0 \alpha}$$

where R_t = resistance at t degrees Celsius

R_0 = resistance at 0°C

α = temperature coefficient of resistance, °C at 0°C

$$R_2 = \frac{R_1(1 + \alpha t_2)}{1 + \alpha t_1}$$

where R_1 = resistance at t_1 degrees Celsius

R_2 = resistance at t_2 degrees Celsius

Cable voltage drop $U = IR$

where R = total cable resistance, Ω

I = cable current, A

$$U_L = U_S - U$$

where U_L = load terminal voltage

U_S = supply voltage

U = cable voltage drop

2.9   Exercises

1 The resistance of 100 m of 4·5 mm^2 cable is 0·36 Ω. What is the resistance of 600 m of this cable?

2 A single-core cable, 24 m long, has a measured conductor resistance of 0·06 Ω. What is the resistance of 1000 m of this cable?

3 The resistance of 15 mm^2 single cable is 1·21 Ω/km. Calculate the resistance of a twin 15 mm^2 cable, 80 m long, if the cores are connected in series.

4 100 m of 40 mm^2 copper cable has a resistance of 0·0445 Ω. What is the resistance of 100 m of 15 mm^2 copper cable?

5 1 km of 4·5 mm^2 copper cable has a resistance of 3·82 Ω. Calculate the resistance of 250 m of 2 mm^2 copper cable.

6 If the resistance of 2 mm^2 copper cable is 8 Ω/km, what length of 1 mm^2 copper cable will have a resistance of 4 Ω?

7 Copper cable of cross-sectional area 40 mm^2 has a resistance of 0·445 Ω/km. What will be the cross-sectional area of a similar cable, 100 m long, which has a resistance of 0·89 Ω?

8 A certain cable has a resistance of 0·5 Ω. What is the resistance of a cable made of the same material which has twice the cross-sectional area and is three times as long as the first?

9 Two cables have equal resistances, but one has a cross-sectional area 2·5 times greater than the other. How much longer is the thicker cable than the thinner cable?

10 In the manufacture of copper wire, a thick circular rod, which has a resistance of 0·01 Ω, is drawn out without change in volume until its diameter is one tenth of what it was. What is its new resistance?

11 (a) How does the electrical resistance of a copper conductor vary with
   (i) an increase in its length
   (ii) an increase in its cross-sectional area
   (iii) an increase in its temperature?
   (b) Place the following materials in ascending order of resistivity:
   glass, copper, iron, carbon, brass.

12 (a) State how the length and the cross-sectional area of a conductor affects its resistance.
   (b) A piece of silver wire and a piece of resistance wire have identical dimensions. State how their electrical resistance will differ, giving the reason for your answer.

13 A twin cable with aluminium conductors with cross-sectional area of 25 mm^2 is 150 m long. What is the resistance of the cable? The resistivity of aluminium is 28·5 μΩmm.

14 Calculate the resistance of 1000 m of copper conductor of 4·5 mm^2 cross-sectional area. Take the resistivity of copper as 17·2 μΩmm.

15 What is the resistance of 200 m of copper conductor with a cross-sectional area of 2 mm^2. The resistivity of copper is 17·2 μΩmm.

16 A cable has a cross-sectional area of 1 mm^2. What length of this cable (to the nearest metre) will have a resistance of 2 Ω? The cable is made of copper with a resistivity of 17·2 μΩmm.

17 An electric-fire element has a total length of 10 m, a cross-sectional area of 0·5 mm^2 and a resistance of 50 Ω. Calculate the resistivity of the material from which it is made.

18 What is the cross-sectional area of a cable having a resistivity of 20 μΩmm if its resistance is 1·5 Ω/km?

19 Calculate the length of a single aluminium conductor of cross-sectional area 125 mm^2 which has a resistance of 0·1 Ω. The resistivity of aluminium is 28·5 μΩmm.

20 500 m of single cable is buried and cannot be examined. The cross-sectional area of the cable is known to be 3 mm^2, and its resistance is measured as $2 \cdot 8$ Ω. Suggest the material from which the conductor is made.

21 The lead sheath of a cable, 600 m long, has a measured resistance of 20 Ω. If the resistivity of lead is 219 μΩmm, find the cross-sectional area of the sheath.

22 The shunt-field winding of a DC motor has a resistance of 100 Ω at 20°C. What will be its resistance at 45°C, if it is made of copper? ($\alpha = 0 \cdot 00396$/°C at 20°C)

23 The resistance of a steel catenary wire is 2 Ω at 20°C, but is used to support lighting fittings in a large coldroom which is kept at a temperature of -10°C. What will be its cold resistance, if α for steel is $0 \cdot 006$/°C at 20°C?

24 The resistance of 1 mm^2 copper conductor is 13 Ω/km at 20°C. What will be its resistance/km at 60°C? ($\alpha = 0 \cdot 00396$/°C at 20°C)

25 A 1 mm^2 MI cable has a resistance of $1 \cdot 5$ Ω at 20°C. When installed to feed control circuits on a furnace, its resistance is measured as $2 \cdot 5$ Ω. What is the average operating temperature? α for copper is $0 \cdot 00396$/°C at 20°C.

26 A resistance thermometer has a temperature coefficient of $0 \cdot 002$/°C at 0°C and a resistance of 40 Ω. When used to measure the temperature of an oven, the resistance increases to 56 Ω. What is the temperature of the oven?

27 The resistance of a coil of cable at 10°C is $3 \cdot 5$ Ω. What will be its resistance at 25°C? The cable conductor is made of aluminium, for which α is $0 \cdot 0040$/°C at 0°C.

28 If a resistance thermometer is made of brass wire and has a resistance of 10 Ω at 15°C, what will be its resistance at 300°C? α for brass is $0 \cdot 0010$/°C at 0°C.

29 A motor is used to drive a conveyor feeding a stove-enamelling plant, and its temperature when idle never falls below 30°C. At this temperature, the winding resistance is measured as 15 Ω. When running, the operating temperature of the motor is 65°C. What is the winding resistance at the higher temperature? The winding is of copper with $\alpha = 0 \cdot 0043$/°C at 0°C.

30 A voltmeter coil has a resistance of 1000 Ω. The meter indicates correctly at 15°C. If the coil temperature rises to 65°C and the temperature coefficient of the coil wire is $0 \cdot 0004$/°C, what reading will it show with 10 V applied?

31 The copper shunt-field coil of a DC motor takes $2 \cdot 5$ A from a 200 V supply when it is first switched on and its temperature is 16°C. What will be the field current when the motor has been running for some time and its temperature increases to 56°C? Take α for copper as $0 \cdot 0043$/°C at 0°C.

32 Calculate the resistance of 200 m of single-core 1 mm^2 MI cable, with hard-drawn copper conductor, at 20°C. What will be the resistance of this cable at 80°C? ρ for hard-drawn copper at 20°C is $17 \cdot 8$ μΩmm. α for hard-drawn copper is $0 \cdot 00396$/°C at 20°C.

33 A resistor used in a measuring circuit has a resistance of 10 kΩ at 20°C, and is made of a material having a temperature coefficient of resistance of

$0\cdot0001/°C$ at 20°C. What is the percentage change in the value of the resistor when its temperature is 30°C?

34 The maximum permissible voltage drop in an installation is 4% of the supply voltage. What will be the minimum load voltage, to the nearest volt, for supply voltages of
(a) 200 V,
(b) 230 V,
(c) 240 V, and
(d) 250 V?

35 Each core of a twin cable carrying a current of 10 A has a resistance of $0\cdot15\ \Omega$. What is the total voltage drop in the cable?

36 An industrial heater is connected to a 240 V supply by means of a twin cable, each core having a resistance of $0\cdot08\ \Omega$. The heater takes a current of 25 A from the supply. What is the PD across its terminals?

37 What must be the maximum resistance of each core of a twin cable, feeding a heater taking 12 A from a 240 V supply, if the allowable voltage drop of 4% of supply voltage must not be exceeded?

38 A laboratory socket outlet is fed through a twin cable with each core having a resistance of $0\cdot2\ \Omega$. The supply is at 115 V, and the cable voltage drop must not exceed $1\cdot5\%$ of this value. What maximum current should be taken from the outlet?

39 A motor is fed from 240 V mains 50 m away by a 2-core cable, each core of which has a cross-sectional area of $8\cdot6\ \text{mm}^2$. Calculate the voltage at the motor terminals when it is taking 30 A. Take the resistivity of copper as $17\cdot2\ \mu\Omega\text{mm}$.

40 An industrial oil heater takes a current of 60 A from a 400 V supply through a twin-core aluminium cable, 80 m long. If the cable voltage drop must not exceed 4% of the supply voltage, calculate the minimum cross-sectional area of each core. Take ρ for aluminium as 28 $\mu\Omega\text{mm}$.

41 An immersion heater takes a current of $12\cdot5$ A and is fed through a twin cable, each core of which has a cross-sectional area of $2\cdot5\ \text{mm}^2$. The cable conductors are made of copper, which has a resistivity of $17\cdot2\ \mu\Omega\text{mm}$. Calculate the greatest length of cable which may be used if the cable voltage drop must not exceed 6 V.

42 A motor takes 10 A from a 200 V supply, and is fed through a twin $2\cdot5\ \text{mm}^2$ copper cable 30 m long. Calculate the voltage at the motor terminals. ρ for copper is $17\cdot2\ \mu\Omega\text{mm}$.

2.10 Multiple-choice exercises

2M1 The volt drop in a cable carrying a direct current is due to its
(a) impedance   (b) resistance   (c) size and type   (d) reactance

2M2   The correct symbol for a fixed resistor is

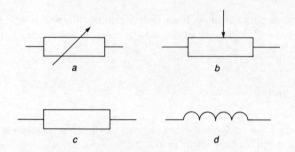

2M3   The resistance of a conductor is
   (*a*) directly proportional to its length
   (*b*) independent of its dimensions
   (*c*) directly proportional to its cross-sectional area
   (*d*) small

2M4   The type of resistor most likely to be found in a radio receiver is a
   (*a*) high-power asbestos woven-mat type
   (*b*) wire-bound encapsulated type
   (*c*) vitreous-enamelled wire-wound type
   (*d*) carbon low-power type

2M5   The resistor shown in the diagram is being used as a

   (*a*) potential divider
   (*b*) fixed-value resistor
   (*c*) heater
   (*d*) rheostat

2M6   The resistance of a certain type of cable is 4 Ω/km. The resistance of 80 m of the same cable will be
   (*a*) 32 Ω (*b*) 50 Ω)   (*c*) 0·32 Ω (*d*) 0·5 Ω

2M7   Resistivity is usually quoted in
   (*a*) ohms   *b*) $\mu\Omega$m   (*c*) microhms   (*d*) rho

2M8 A conductor has a cross-sectional area of 20 mm^2, a length of 48 m and is made of a material with a resistivity of 0·02 μΩm. Its resistance is
(*a*) 48 kΩ   (*b*) 48 Ω   (*c*) 0·48 Ω   (*d*) 0·048 Ω

2M9 If a carbon resistor increases in temperature its resistance will
(*a*) decrease   (*b*) be unchanged   (*c*) increase   (*d*) fall to zero

2M10 The temperature coefficient of resistance of a conductor
(*a*) indicates the maximum temperature that a resistor may reach
(*b*) allows the calculation of resistance as temperature changes
(*c*) shows that a resistor will get hot when it carries current
(*d*) shows that its resistance will fall as it gets hotter

2M11 The current to a tungsten-filament lamp at the moment of switch-on
(*a*) will blow the circuit fuse
(*b*) is lower than the steady running current
(*c*) usually causes the filament to burn out
(*d*) is several times greater than after a second or so

2M12 The maximum voltage drop permitted in a circuit by *BS 7671* is
(*a*) zero   (*b*) 0·4%   (*c*) 4%   (*d*) 12 V

Chapter 3
Mechanics

At first sight, it may seem strange that a book dealing with electrical theory should concern itself with mechanics. What connection is there?

As well as understanding the operation of electrical equipment, an electrical craftsman has to manufacture and install it. This involves various mechanical operations, such as cutting cables, threading conduits, driving screws, lifting heavy apparatus, and so on. An understanding of the principles involved can hardly fail to reduce both the physical effort required, and the likelihood of accident, in carrying out these tasks.

3.1 Mass, force, pressure and torque

Mass

Mass can be defined as the amount of material in an object, and is usually measured by comparison with another mass chosen as a standard. For instance, the standard of mass in SI units is the **kilogram** (kg), which is the mass of a block of platinum kept at Sèvres in France.

Force

A scientific definition of force is difficult at this stage, but it will suffice to say that force can be measured in terms of the effects it produces. A force can lift, bend or break an object. It can move an object previously stationary, or can stop or change the rate and direction of movement. In electrical-craft practice, force is used to move, bend, cut or join together materials, as well as to drive in screws and perform a host of other tasks.

Weight

The Earth exerts a natural force of attraction on all other masses. This force is generally called the 'gravitational pull' of the Earth on a body, and the magnitude of this force measures the weight of the body.

Units of force and weight

A mass of one kilogram will experience a force due to the Earth's gravity of one kilogram force. However, since the kilogram is a unit of mass, it should not be used to measure force. The SI unit of force is the **newton** (symbol N). A mass of one kilogram experiences a force due to gravity of 9·81 newtons.

We could say, for example, that a piece of machinery has a mass of 1 kg, in which case it will require a force of 9·8 N to lift it against gravity. If the machinery formed part of a space rocket, it could be sent outside the gravitational pull of the Earth. Its mass would then be unchanged, but it would experience no Earth gravitational force.

Example 3.1

A bundle of conduit has a mass of 800 kg. What force will be needed to lift it?

One mass of 1 kg requires a force of 9·8 N to lift it against gravity.

$$\text{force required } = 800 \times 9 \cdot 81 \text{ newtons}$$
$$= 7848 \text{ N}$$

Pressure

The pressure on a surface is measured in terms of **the force per unit area on the surface**, assuming that the surface is at right angles to the direction of the force. For example, a force of 500 N acting on one leg of a tripod pipevice with a sharp edge of effective area 20 mm^2 in contact with a floor will exert a pressure given by

$$\text{pressure} = \frac{\text{force}}{\text{effective area}} = \frac{500 \text{ N}}{20 \text{ mm}^2} = 25 \text{ N/mm}^2$$

A pressure as high as this will damage many floor finishes. If the leg of the vice is stood on a piece of wood 100 mm square, the same force is spread over an area of 10 000 mm^2, and the pressure falls to

$$\frac{500 \text{ N}}{10\ 000 \text{ mm}^2} = 0 \cdot 05 \text{ N/mm}^2$$

Strictly speaking, the SI unit of pressure is the newton per square metre or pascal. Since there are 1 000 000 mm^2 in 1 m^2, the pressure of 0·05 N/mm^2 given above could also be expressed as 5×10^4 N/m^2.

Density

If we take two blocks of the same size, one made of wood and the other of iron, the iron block will be heavier than the wooden block. Thus we can say that iron is more dense than wood. The density of a material is the mass of it which is contained in one cubic unit, and is given in kilograms per cubic metre (kg/m^3).

Example 3.2

What will be the mass of a block of copper $0 \cdot 1$ m by $0 \cdot 2$ m by $0 \cdot 15$ m? The density of copper is 8900 kg/m^3.

The volume of the block is

$$0 \cdot 1 \times 0 \cdot 2 \times 0 \cdot 15 = 0 \cdot 003 \text{ m}^3$$

$$\text{since density} = \frac{\text{mass}}{\text{volume}}$$

$$\text{mass} = \text{density} \times \text{volume}$$

$$= 8900 \times 0 \cdot 003 \text{ kilograms}$$

$$= 26 \cdot 7 \text{ kg}$$

Torque

In some cases, a force may tend to cause rotary movement. Probably the simplest example of this principle in electrical-craft work is the process of threading a conduit with a stock and die. The turning effect is called the **torque**, or sometimes the **turning moment of a force**, and is measured as the force applied multiplied by the perpendicular distance between the direction of the force and the point about which rotation can occur. Figure 3.1 makes it clear that, to obtain maximum torque for a given applied force, the force should always be in the direction of the resulting movement.

Torque (or turning moment) is measured in newton metres (Nm). In Figure 3.1*a*, there is no torque; although there is a force of 100 N, this is trying to compress the stock and not to rotate it, since the distance between the direction of the force and the turning point is zero. The torque produced in Figure 3.1*b* is less than that in Figure 3.1*c*. The forces are the same in both cases, but, in the latter example, the distance between the direction of the force and the turning point is maximum. In other words, all the force applied is being used to turn the stock, and none is wasted in trying to distort the stock. The actual torque required to thread a conduit depends on its size, but can be reduced considerably by the use of a sharp die and correct lubrication. For a given torque, the force required can be reduced if exerted at a greater radius from the turning point (Example 3.3).

Example 3.3

A conduit requires a torque of 20 Nm for the threading operation. What minimum force must be exerted by each hand pressing at the ends of a stock of overall length (*a*) $0 \cdot 5$ m, (*b*) 1 m?

Each hand provides half the total torque, so the torque per hand will be 10 Nm.

$$\text{since torque} = \text{force} \times \text{direct distance}$$

$$\text{force} = \frac{\text{torque}}{\text{direct distance}}$$

In both cases, the direct distance from the point of application of the force to the turning point will be half the overall length of the stock.

(*a*) minimum force required $= \dfrac{10 \text{ Nm}}{0 \cdot 25 \text{ m}} = 40 \text{ N}$

(*b*) minimum force required $= \dfrac{10 \text{ Nm}}{0 \cdot 5 \text{ m}} = 20 \text{ N}$

Figure 3.1   Torque applied to stocks and dies
a Torque $= 100 \text{ N} \times 0 \times 2 = \text{zero}$
b Torque $= 100 \text{ N} \times 0 \cdot 1 \text{ m} \times 2 = 20 \text{ Nm}$
c Torque $= 100 \text{ N} \times 0 \cdot 3 \text{ m} \times 2 = 60 \text{ Nm}$

The lesson drawn from this example should not tempt the technician to increase the effective length of their tool handles by the use of a suitable length of pipe or other means. The excessive torque that can then be applied is often sufficient to damage tools. A tool will usually be made with handles of such length as to prevent the application of too much torque by a person of normal strength.

3.2 Work, energy and power

Work

If a force is applied to a body and movement results, work is done. For instance, work is done when a weight is lifted or when forces applied to a stock cause it to rotate.

Work is measured as the product of the distance moved by the object and the force which caused the movement. When the movement is in the same direction as the force, the work done is equal to the distance moved multiplied by the force exerted.

$$\text{work} = \text{distance} \times \text{force}$$

The unit is the **metre newton** (mN), which is also known as the **joule** (J).

Example 3.4

A bundle of conduit has a mass of 200 kg. What is its weight? What work must be done in lifting the conduit from the floor on to a rack 2 m high?

Weight is the force exerted on the mass by gravity. Since a mass of 1 kg exerts a force due to gravity of $9 \cdot 81$ N,

$$\text{weight} = \text{mass} \times 9 \cdot 81 \text{ newtons}$$
$$= 200 \times 9 \cdot 81 \text{ newtons}$$
$$= 1962 \text{ N}$$
$$\text{work done} = \text{distance} \times \text{force}$$
$$= 2 \text{ m} \times 1962 \text{ N}$$
$$= 3924 \text{ mN or } 3924 \text{ J}$$

Example 3.5

A force of 100 N will just move a van on a level road. What work will be expended in pushing the van 15 m?

$$\text{work} = \text{distance} \times \text{force}$$
$$= 15 \text{ m} \times 100 \text{ N}$$
$$= 1500 \text{ mN or } 1500 \text{ J}$$

Since the van is not being lifted, its weight is only important in its effect on the force needed to move it against friction.

Energy

Energy is the capacity to do work. It may take many forms, such as nuclear, chemical, thermal, mechanical or electrical energy. If we ignore the theoretical atomic physics, which never affects electrical-craft work, it is true to say that, whereas energy can be converted from one form to another, it can be neither created nor destroyed. For instance, coal or oil, containing chemical energy, is burned in the boilers of a power station, and produces heat. This heat evaporates water to become steam under pressure, which is fed to a turbine where mechanical energy is produced. The turbine drives an alternator which produces electrical energy.

The units of energy are the same as those of the work it is capable of performing, i.e. the metre newton (joule).

Efficiency

Although energy cannot be created or destroyed, it does not follow that energy can be converted from one form to another without waste. For instance, the conversions in a power station generally result in little more than one-third of the available chemical energy becoming electrical energy. The difference is not destroyed, but is dissipated, largely in the form of heat, and is lost from the process. Improvements in design have reduced these losses considerably in recent years. Thus we can say that power stations have become more efficient, because, for a given input of energy, the output energy is greater.

The efficiency of a mechanical system can be defined as the ratio of output to input energy or power, and is usually expressed as a percentage, so that

$$\text{efficiency} = \frac{\text{output energy or power}}{\text{input energy or power}} \times 100\%$$

In practice, it is more usual to express efficiency in terms of power than energy.

The difference between input and output is the wasted energy, or losses, so that

$$\text{output} = \text{input} - \text{losses}$$

$$\text{input} = \text{output} + \text{losses}$$

Hence two further ways of defining efficiency follow:

$$\text{efficiency} = \frac{\text{output}}{\text{output} + \text{losses}} \times 100\%$$

$$\text{efficiency} = \frac{\text{input} - \text{losses}}{\text{input}} \times 100\%$$

Power

Power is defined as the rate of doing work. An electrician using a hammer and chisel can pierce a hole in a brick wall. The same hole could be produced more quickly with a pneumatic drill. In either method, the completed hole will represent the same amount of work, but the rate of doing work is greater in the second method because the pneumatic drill is more **powerful** than the electrician. Thus

$$\text{average power} = \frac{\text{amount of work done}}{\text{time taken to do it}}$$

The unit is the **joule per second** (J/s), which is also known as the **watt** (W). The watt is too small for many practical applications, so the kilowatt (kW) is often used.

$$1 \text{ kW} = 1000 \text{ W} = 1000 \text{ J/s}$$

Example 3.6

An electric motor drives a pump which lifts 1000 l (litres) of water each minute to a tank 20 m above normal water level. What power must the motor provide if the pump is 50% efficient? One litre of water weighs 9·81 N.

$$\text{rate of lifting water} = 1000 \times 9 \cdot 81 = 9810 \text{ N/min}$$

$$\text{power required by pump} = 9810 \times 20 = 196\ 200 \text{ mN/min}$$

$$= \frac{196\ 200}{60} \text{ newton metres per second (watts)}$$

$$= 3270 \text{ W}$$

$$\text{output power of motor} = 3270 \times \frac{100}{50} \text{ W}$$

$$= 6540 \text{ W or } 6 \cdot 54 \text{ kW}$$

3.3   Lifting machines

A lifting machine is a device which can enable a small effort to lift a large load, and such machines are often necessary in electrical work to move heavy equipment. A few of the simpler types are considered here.

The lever

Consider the simple lever (Figure 3.2). The total turning moment or torque applied to one end of the lever must be equal and opposite to the torque available at the other end. Thus, if a 50 N effort is applied to the right-hand end of the lever as shown, the total applied torque will be 50 N × 2 m, or 100 Nm. If there are no torque losses (and losses here will be very small)

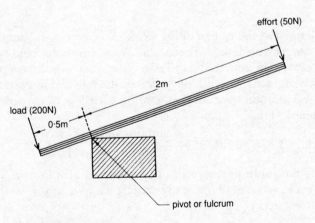

Figure 3.2   Simple lever

the same torque applied to the right-hand end of the lever will be available at the left-hand end. Since the radius of action about the pivot, or fulcrum, is $0 \cdot 5$ m at this end,

$$100 \text{ Nm} = 0 \cdot 5 \text{ m} \times \text{available force}$$

$$\text{available force} = \frac{100 \text{ Nm}}{0 \cdot 5 \text{ m}} = 200 \text{ N}$$

This shows that this lever enables a load of 200 N to be balanced by a force of 50 N, because the lever is four times as long on one side of the fulcrum as it is on the other.

Levers are classified according to the relative positions of the load, the effort and the fulcrum, as shown in Figure 3.3.

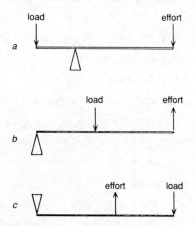

Figure 3.3   Classification of simple levers
    a Class 1 lever
    b Class 2 lever
    c Class 3 lever

Mechanical advantage and velocity ratio

The lever of Figure 3.2 has the advantage of allowing a large mass to be lifted by a small force, and mechanical advantage is the ratio of load to effort.

$$\text{mechanical advantage (MA)} = \frac{\text{load}}{\text{effort}}$$

For the lever of Figure 3.2, for example,

$$\text{MA} = \frac{\text{load}}{\text{effort}} = \frac{200 \text{ N}}{50 \text{ N}} = 4$$

This is a ratio, so there is no unit.

For a simple lever, mechanical advantage will be almost constant and, for a class 1 lever, will depend on the lengths of the lever on either side of the fulcrum. For most machines, however, frictional waste will vary with load, and mechanical advantage will not be constant.

Velocity ratio is defined as the ratio of the distances moved by the effort and the load:

$$\text{velocity ratio (VR)} = \frac{\text{the distance moved by the effort in a given time}}{\text{the corresponding distance moved by the load}}$$

For example, it will be clear from Figure 3.2 that, if the load moved by 1 mm, the effort must move by 4 mm, so that the velocity ratio will be 4 (no units apply).

In an ideal machine (that is, a machine which is 100% efficient),

$$\text{work input} = \text{work output}$$

so

$$\text{effort} \times \text{distance moved by effort} = \text{load} \times \text{distance moved by load}$$

$$\frac{\text{distance moved by effort}}{\text{distance moved by load}} = \frac{\text{load}}{\text{effort}}$$

$$\text{velocity ratio} = \text{mechanical advantage}$$

For most machines, the inefficiency results in a smaller value for mechanical advantage than for velocity ratio, and it is interesting to note that

$$\text{efficiency} = \frac{\text{work out}}{\text{work in}}$$

$$= \frac{\text{load} \times \text{distance moved by load}}{\text{effort} \times \text{distance moved by effort}}$$

$$= \frac{\text{load}}{\text{effort}} \div \frac{\text{distance moved by effort}}{\text{distance moved by load}}$$

$$= \frac{\text{mechanical advantage}}{\text{velocity ratio}}$$

Example 3.7

A class 1 lever is arranged so that a load of 1000 N is to be lifted at a distance of 200 mm from the fulcrum. What force must be applied 2 m from the fulcrum to balance the load? Calculate the mechanical advantage and velocity ratio of the system, assuming no losses.

Let F be the required force.

$$\text{turning moment required for load} = \frac{200}{1000} \times 1000 \text{ newton metres}$$

$$= 200 \text{ Nm}$$

If there is no waste, this will equal the turning moment to be applied.

Therefore $F \times 2 = 200$ Nm

and $$F = \frac{200 \text{ Nm}}{2 \text{ m}} = 100 \text{ N}$$

$$\text{MA} = \frac{\text{load}}{\text{effort}} = \frac{1000 \text{ N}}{100 \text{ N}} = 10$$

$$\text{VR} = \frac{\text{effort distance}}{\text{load distance}} = \frac{2 \text{ m}}{0 \cdot 2 \text{ m}} = 10$$

Inclined planes and jacks

A simple method of lifting a heavy load is to push it up an inclined plane. For example, heavy machinery can be manhandled on to a lorry with comparative ease if a number of planks form an inclined plane from the ground to the lorry. If necessary, a packing case or trestle can be placed beneath the centre of the planks to give extra support. The load is pushed up the sloping planks in a series of stages. This will involve more work than a direct lift, owing to the friction between the load and the planks, but the force required at any instant is less and rests may be taken between efforts. A useful safety measure to ensure that the load does not slip can be taken by passing a rope, secured at one end to the load, round a solidly fixed structure such as a girder or the lorry chassis. Slack rope is taken up as the load is lifted.

Figure 3.4   Simple jack

In some machines, a steel inclined plane is, in effect, wound round a central column to form a screw thread. Rotation of the thread can lift a load bearing on it by forcing it up the inclined plane. A machine of this sort is called a jack. One type of jack is shown in Figure 3.4.

Pulley systems

A machine which is often used to lift heavy loads where a suitable overhead suspension is available is the 'block and tackle'. This consists of two sets of pulleys and a length of rope. A simple arrangement with two blocks each having two pulleys is shown in Figure 3.5. The pulleys in each block are usually the same size and are mounted side by side, but are drawn as shown for clarity.

Neglecting the rope to which the effort is applied, there are four ropes supporting the load. If the load is to be lifted by, say, one metre, each of these four ropes must shorten by one metre, so that the effort rope must be pulled through four metres. The velocity ratio is thus equal to four in this case, and, in fact, the velocity ratio of a block and tackle is equal to the number of pulleys used. The mechanical advantage (MA) would be equal to the velocity ratio (VR) if there were no losses, but friction at the pulley bearings reduces its value.

effort

load

Figure 3.5   Principle of block and tackle

Example 3.8

A block and tackle of two sets, each of five pulleys, is 60% efficient. What is the maximum load which a man weighing 600 N could support using it?

$$\text{VR} = 2 \times 5 = 10$$

$$\text{MA} = \text{VR} \times \text{efficiency} = 10 \times \frac{60}{100} = 6$$

Maximum effort will be applied when the man hangs on the rope. This is an effort of 600 N.

since
$$\text{MA} = \frac{\text{load}}{\text{effort}}$$

$$\text{load} = \text{MA} \times \text{effort} = 6 \times 600 \ \text{newtons} = 3600 \ \text{N}$$

3.4 Power transmission

Belts and pulleys

A method of connecting a motor to a machine is to fit both motor and machine with suitable pulleys, the two connected by a belt. Most belts are of the 'vee' type, which grip the pulley and reduce slip.

If the two pulleys are the same size, speeds and torques will be identical if belt slip is neglected. If the pulleys are of differing sizes, a speed change can be arranged. Consider the arrangement shown in Figure 3.6 where a motor of speed N_1 revolutions per minute is fitted with a pulley of diameter D_1 metres. The motor pulley is coupled by a belt to a machine pulley of diameter D_2 metres and having a speed of N_2 revolutions per minute.

The belt makes contact with both pulleys; so, if we neglect belt slip, the edge speed, or peripheral speed, of each pulley is the same. The motor pulley has a circumference of πD_1 metres, and, since it makes N_1 revolutions each minute, its peripheral speed is $\pi D_1 N_1$ metres per minute. Similarly, the peripheral speed of the machine pulley is $\pi D_2 N_2$ metres per

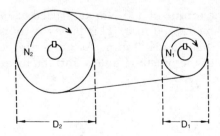

Figure 3.6   Belt and pulleys. Two pulleys coupled by belt

minute. Since the peripheral speeds of both pulleys are equal if belt slip is neglected,

$$\pi D_1 N_1 = \pi D_2 N_2$$

so that
$$\frac{N_1}{N_2} = \frac{D_2}{D_1}$$

Thus pulley speeds are in inverse ratio to their diameters.

As well as changing the speed, pulleys of differing sizes also change the torque. It would be wrong to state that no losses occurred in the belt-drive system; but, if we neglect these losses, the work done by the motor is provided to the machine, so that the two are equal. Let the force exerted at the edge of the motor pulley be F_1 newtons, and that at the machine pulley be F_2 newtons. In one revolution, the force on the motor pulley moves through πD_1 metres, so that the work done in 1 min is $\pi D_1 N_1 F_1$ joules. Similarly, the work done on the machine pulley in 1 min is $\pi D_2 N_2 F_2$ joules.

Assuming no losses, these values are the same, so that

$$\pi D_2 N_2 F_2 = \pi D_1 N_1 F_1$$

or
$$D_2 F_2 N_2 = D_1 F_1 N_1$$

The torque on the motor pulley is $F_1(D_1/2)$, and that on the machine pulley is $F_2(D_2/2)$.

If motor torque is T_1, and machine torque is T_2,

$$T_2 N_2 = T_1 N_1$$

or
$$\frac{T_1}{T_2} = \frac{N_2}{N_1} = \frac{D_1}{D_2}$$

Thus pulley torques are in inverse ratio to their speeds, but in direct ratio to their diameters. If a belt-and-pulley system is used to reduce speed, it must increase torque. Similarly, a pulley-and-belt system intended to increase speed will reduce available torque.

Example 3.9

A motor providing a torque of 300 Nm at a speed of 1440 r/min is fitted with a pulley of diameter 100 mm. This pulley is coupled by a belt to a machine pulley of diameter 400 mm. Assuming no losses or belt slip, calculate the speed of the machine pulley and the torque it can provide.

$$\frac{N_1}{N_2} = \frac{D_2}{D_1}$$

Therefore
$$N_2 = \frac{N_1 D_1}{D_2}$$

$$= 1440 \times \frac{100}{400} \text{ revolutions per minute}$$

$$= 360 \text{ r/min}$$

Also,
$$\frac{T_1}{T_2} = \frac{D_1}{D_2}$$

Therefore
$$T_2 = \frac{T_1 D_2}{D_1}$$

$$= 300 \times \frac{400}{100} \text{ newton metres}$$

$$= 1200 \text{ Nm}$$

Gearing

Instead of using a belt to connect two pulley wheels, each may have teeth cut in its edge. If the teeth are enmeshed, they must turn together and have the same peripheral speed. Toothed wheels of this sort are called **gears**. They are more positive than a belt as they cannot slip, but may be noisy, and will usually require periodic lubrication.

Since peripheral speeds are the same, the same reasoning applies to gears as to pulleys. It should, however, be noted that two pulleys coupled by a belt rotate in the same direction, whereas two meshed gears rotate in opposite directions. Gear sizes are usually indicated by the number of teeth rather than by diameter, the two being proportional. Thus, if G_1 and G_2 represent the numbers of teeth on two gearwheels, respectively,

$$\frac{T_1}{T_2} = \frac{N_2}{N_1} = \frac{G_1}{G_2}$$

Example 3.10

A motor with a speed of 720 r/min provides a torque of 50 Nm, and is fitted with a gearwheel having 30 teeth. This wheel is meshed with a second gearwheel, which has 12 teeth. Calculate the speed of the second gearwheel, and the torque it provides.

$$N_2 = \frac{N_1 G_1}{G_2} = 720 \times \frac{30}{12} \text{ revolutions per minute} = 1800 \text{ r/min}$$

$$T_2 = \frac{T_1 G_2}{G_1} = 50 \times \frac{12}{30} \text{ newton metres} = 20 \text{ Nm}$$

Chains

A chain coupled to two or more rotating sprockets can be considered as a pulley-and-belt system, and follows the same rules. The chain is more expensive and is noisier than the belt, but cannot slip. It is used where a positive drive is needed.

Rotating shafts

As well as the drives already considered, the simple rotating shaft is a common mechanical power transmitter. Without belts, gears or chains, there is no change in speed or in torque with a shaft drive.

3.5   Parallelogram and triangle of forces

Scalars and vectors

Mechanical-engineering quantities can be classified as either scalars or vectors. Those which can be measured in terms of magnitude only are called **scalar** quantities. Examples of scalar quantities are length, mass and power.

Other quantities can only be completely described if reference is made to direction as well as to magnitude, and these are called **vector** quantities. For instance, the position of an aircraft after a given time can only be forecast accurately if not only its starting position and speed, but also its direction, are known.

A force is a vector quantity and is often represented by a line called a **vector**. The length of the vector, drawn to a suitable scale, represents the magnitude of the force, and the direction of the line is the same as that of the force it represents. An arrowhead is added to show the sense (push or pull) of the force, which is assumed to act at a point at one end of the vector.

Equilibrium

A body is said to be in a state of equilibrium when it is at rest or in a state of uniform motion. For instance, a lighting fitting suspended from a chain attached to a ceiling is in equilibrium because it does not move. The downward acting force due to the weight of the fitting is opposed by the equal upward supporting force of the chain. If the weight of the fitting is too great for its fixings to bear, the downward force exceeds the upward force, equilibrium is lost and the fitting will fall.

Resultant of forces

It often happens that a body is acted on simultaneously by a number of forces. For simplicity, we can assume that these forces are removed and replaced by one force, called the **resultant**, which will have exactly the same effect on the body as the forces which it replaces. The resultant of two forces can be found by completing the **parallelogram of forces** as shown in Figure 3.7. Two forces P and Q, represented by the vectors OA and OB, have the resultant R, represented by the vector OC. If a resultant

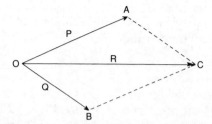

Figure 3.7   Parallelogram of forces. *R* is resultant of forces *P* and *Q*

for more than two forces is required, a similar procedure is adopted, taking the resultant of two vectors at a time until the system is resolved into one vector only.

It is often necessary to find the force which will just balance all the other forces acting on a body. This is called the **equilibrant**, and is equal in magnitude and opposite in direction to the resultant of the forces.

Example 3.11

A pole supporting overhead cables is situated so that the cables meet it at a right angle. The cables in one direction exert a pull of 4000 N on the pole, and those in the other direction a pull of 3000 N. Find the horizontal pull on a stay wire, and the direction in which it must be fixed so as to balance the horizontal forces exerted by the cables on the pole.

The arrangement of the system is shown in Figure 3.8. O represents the pole, and vectors OA and OB represent the forces of 4000 N and 3000 N, respectively. These vectors are drawn to scale and at right angles to each other. The parallelogram of forces is now completed, the resultant vector OC representing a force of 5000 N in the direction shown.

The stay wire must be the equilibrant of the resultant OC, and is represented by the equal and opposite vector OD. Thus the horizontal pull

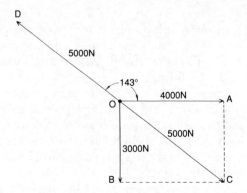

Figure 3.8   Vector diagram for Example 3.11

on the stay wire will be 5000 N, and its direction, measured on the vector diagram with a protractor, will be approximately 143° from the 4000 N cables in an anticlockwise direction.

Triangle of forces

In Figure 3.9*a*, the parallelogram of forces is used to find the resultant *C* and the equilibrant *D* of two forces *A* and *B*. Figure 3.9*b* shows a part of the parallelogram used to find the resultant *C*, and illustrates a useful saving in space which could lead to a larger scale and improved accuracy. Figure 3.9*c* is the same as Figure 3.9*b*, but with the equilibrant *D* substituted for the resultant *C*. The three forces shown form the three sides of the triangle of forces.

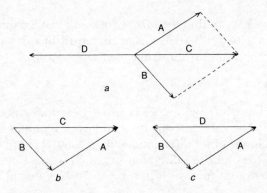

Figure 3.9   Triangle of forces

    a Parallelogram of forces used to find resultant *C* and equilibrant *D* of two forces *A* and *B*

    b Part of parallelogram of forces used to find resultant of same two forces

    c Triangle of forces used to find equilibrant of same two forces

Example 3.12

A fluorescent light fitting weighing 140 N is to be suspended from a single support chain connected to two sling chains taken to the fitting ends as shown in Figure 3.10*a*. Find the tension in each sling chain, if each makes an angle of 30° with the fitting.

A simple figure, often called a space diagram (Figure 3.10*b*) will help us to understand the problem. Since the fitting is in a state of equilibrium, so must be the three forces in the space diagram, where OA is the tension in the support chain, and OB and OC are the tensions in the two sling chains. Clearly, the tension in the support chain must be equal to the weight of the fitting, and will be 140 N.

The triangle of forces is now drawn. First OA, to a scale length equivalent to 140 N, is drawn as shown in Figure 3.10*c*. Next, lines are drawn from

Figure 3.10 Space diagram and triangle of forces of Example 3.12
 a Fluorescent light fitting supported from sling chains
 b Space diagram for arrangement of Figure 3.10*a*
 c Triangle of forces for arrangement of Figure 3.10*a*

points O and A, each making an angle of 60° with OA and meeting at point D to complete the triangle. These lines represent the tensions in the sling chains, and, when measured, are found to have a length which is equivalent to 140 N.

Although the sling chains support the weight between them, each has the same tension as the support chain. Failure to appreciate the fact that the tension in the sling chains increases as the angle they make with the fitting decreases has been responsible for a number of accidents. Here the sling angles are equal; if they were unequal, the chain tensions would be different.

3.6   Summary of formulas for Chapter 3

A mass of 1 kg experiences a force due to gravity of 9·81 N.

$$\text{pressure, N/m}^2 = \frac{\text{force N}}{\text{effective area, m}^2}$$

$$\text{density, kg/m}^3 = \frac{\text{mass, kg}}{\text{volume, m}^3}$$

$$\text{torque, Nm} = \text{turning force, N} \times \text{radius of action, m}$$

$$\text{work or energy, J} = \text{force required, N} \times \text{distance moved, m}$$

$$\text{efficiency} = \frac{\text{output}}{\text{input}} = \frac{\text{output}}{\text{output} + \text{losses}} = \frac{\text{input} - \text{losses}}{\text{input}}$$

$$\text{power, W} = \frac{\text{work or energy, J}}{\text{time, s}}$$

$$\text{mechanical advantage} = \frac{\text{load, N}}{\text{effort, N}}$$

$$\text{velocity ratio} = \frac{\text{distance moved by effort, m}}{\text{distance moved by load, m}}$$

$$\text{efficiency} = \frac{\text{mechanical advantage}}{\text{velocity ratio}}$$

$$\frac{N_1}{N_2} = \frac{D_2}{D_1} = \frac{T_2}{T_1}$$

where N_1 is the speed of the first pulley or gear, r/min
N_2 is the speed of the second pulley or gear, r/min
D_1 is the diameter of the first pulley or gear, m
D_2 is the diameter of the second pulley or gear, m
T_1 is the torque provided to, or by, the first pulley or gear, Nm
T_2 is the torque provided to, or by, the second pulley or gear, Nm

3.7   Exercises

1 A sling is marked as having a safe working load of 2000 kg. What weight will it support safely?

2 A machine requires a force of 500 N to lift it. What is its mass?

3 A masonry drill has an effective surface area of 50 mm^2 and, for best rate of penetration, must be operated at a pressure of 4 N/mm^2. What force should be applied to the drill?

4 An apprentice weighs 480 N. What pressure will be exerted on a plasterboard ceiling if he stands between the joists
(*a*) on one foot which has an area of contact with the ceiling of 8000 mm^2
(*b*) on a rigid plank 300 mm wide and 2 m long?

5 A turning moment of 18 Nm is required to tighten a nut on a busbar clamp. What maximum force must be applied to a spanner of effective length $0 \cdot 27$ m if the nut is not to be overtightened?

6 An electrician exerts a turning force of 50 N on each of two handles of a set of stocks and dies. The effective length of each handle is 200 mm. What total turning moment is applied to the dies?

7 How much energy must be expended to raise a bundle of conduit weighing 600 N from a floor to a scaffolding 7 m above it? If the task takes two minutes to complete, what average power is used?

8 A motor has timbers bolted to it for protection, and requires a horizontal force of 450 N to move it over a level floor. How much work is done in moving it 50 m?

9 The petrol engine of a builder's hoist is to be replaced by an electric motor. What is the power rating of the motor if it must be capable of lifting 2400 N through 32 m in 24 seconds, the hoist gear being 80% efficient?

10 The pivot of a hydraulic pump is at one end of the handle, which is 1·2 m long, and is 0·2 m from the attachment to the pump rod. The pump piston has an effective surface area of 200 mm². If the handle of the pump is pushed down by a force of 80 N, calculate (*a*) the force on the pump rod and (*b*) the pressure on the pump piston.

11 What must be the minimum weight of a man who is able to lift a load of 5600 N using a crowbar 2 m long, with the fulcrum 0·2 m from the load end, and what is the mechanical advantage?

12 A block and tackle of eight pulleys is to be used to lift a cubicle switch panel weighing 2400 N. The lifting system is 80% efficient. Calculate the effort required.

13 A screw jack has a velocity ratio of 112 and an efficiency of 50%. What force must be applied to it just to lift a drum of cable weighing 8960 N?

14 A load pulley is 0·5 m in diameter, turns at 200 r/min, and has a torque of 800 Nm. If the motor speed is 720 r/min, calculate the size of the motor pulley and the torque it provides.

15 A load shaft is required to turn at 725 r/min and to provide a torque of 1000 Nm. If the pulley on the 1450 r/min motor is 250 mm in diameter, calculate the diameter of the load pulley and the torque provided by the motor.

16 A motor providing a torque of 100 Nm at 2900 r/min is fitted with a gearwheel having 20 teeth. A drive is to be provided to a shaft having a gearwheel in mesh with that on the motor. If the shaft must provide a torque of 650 Nm, how many teeth has the driven gear, and what is its speed?

17 Two horizontal cables are attached to a pole. The first exerts a force of 2000 N, and the second a force of 3200 N making an angle of 120° with the first. Find the resultant pull on the pole, and its direction with respect to the first cable.

18 Two runs of cable exert forces of 6000 N and 8000 N, respectively, on an overhead-line tower to which they are attached at right angles. What must be the direction of a stay, and what will be the horizontal force it supports, if there is to be no resultant horizontal force on the tower?

19 A set of pulley blocks, having four pulleys in the top block and three ulleys in the lower block, is fixed to a ceiling beam in a workshop. It is to be used to lift a motor weighing 2520 N from its bedplate. The efficiency of the tackle is 60%. Calculate the pull required on the free end of the rope to raise the motor. Make a diagrammatic sketch of the tackle in use.

20 A motor armature weighing 2000 N is freely suspended from a crane hook by means of a double sling with 1 m chains. The motor shaft is horizontal, and the slings are attached to the motor shaft 1 m apart.

(*a*) Draw a diagram showing the arrangement.

(*b*) Determine the tension in the sling.

(*c*) What would be the tension if the chains were 0·835 m instead of 1 m long?

3.8   Multiple-choice exercises

3M1   The SI unit of mass is the
(*a*) newton   (*b*) weight   (*c*) kilogram   (*d*) force

3M2   A man exerts a force due to gravity of 750 N when he stands on one foot which is subject to a pressure of 30 kN/m^2. What is the area of his foot?
(*a*) 250 cm^2 (*b*) 40 cm^2 (*c*) 25 cm^2 (*d*) 2·5 m^2

3M3   The density of a piece of material can be defined as
(*a*) how heavy it is
(*b*) how thick it is
(*c*) if it floats in water
(*d*) the mass of one cubic metre of it

3M4   The unit of torque, the turning moment of a force, is the
(*a*) newton metres (Nm)   (*b*) angular twist in degrees
(*c*) force on a lever   (*d*) metre newtons (mN)

3M5   Work (or energy) can be defined as
(*a*) something hard
(*b*) force exerted times distance moved
(*c*) power in watts
(*d*) the force required to turn a spanner

3M6   The unit of energy is the
(*a*) watt   (*b*) kilogram   (*c*) newton metre   (*d*) joule

3M7   Efficiency can be defined as
(*a*) $\dfrac{\text{input}}{\text{output}}$ (*b*) $\dfrac{\text{output} - \text{losses}}{\text{output}}$

(*c*) $\dfrac{\text{output}}{\text{output} + \text{losses}}$ (*d*) $\dfrac{\text{output} + \text{losses}}{\text{losses}}$

3M8   An electric motor drives a pump which lifts 12 litres (118 N) of water to a height of 8 m. If the combined efficiency of the motor and the pump is 35%, the motor input power will be
(*a*) 330 W   (*b*) 2·70 kW   (*c*) 274 W   (*d*) 944 W

3M9   A class 1 lever is 2·4 m long and is used to lift a load weighing 2000 N at a distance of 0·4 m from the fulcrum. The effort required will be
(*a*) 400 N   (*b*) 333 N   (*c*) 12 000 N   (*d*) 10 000 N

3M10 A jack used to lift heavy loads is
  (*a*) the name of an extremely strong man
  (*b*) a form of class 2 lever
  (*c*) a complicated form of the block and tackle
  (*d*) an inclined plane wound round a central column to form a screw thread

3M11 A block and tackle has a total of six pulleys and is 75% efficient overall. If the maximum possible effort on the operating rope is 200 N, it will be possible to lift a load with a maximum weight of:
  (*a*) 1200 N   (*b*) 1600 N   (*c*) 25 N   (*d*) 900 N

3M12 A motor with a speed of 18 r/s in a clockwise direction is fitted with a vee belt pulley of diameter 20 cm. If the pulley is connected to a drive pulley of diameter 30 cm, the speed and direction of the driven pulley will be:
  (*a*) 27 r/s anticlockwise   (*b*) 12 r/s clockwise
  (*c*) 27 r/s clockwise   (*d*) 12 r/s anticlockwise

3M13 Two gears have 24 teeth and 120 teeth, respectively, and are meshed together. If the output is taken from the 120-tooth gear, which provides a torque of 15 Nm at a speed of 200 r/min, the speed and torque of the 24-tooth driven gear, assuming no losses, will be
  (*a*) 1000 r/min, 3 Nm   (*b*) 500 r/min, 75 Nm
  (*c*) 40 r/min, 75 Nm   (*d*) 1000 r/min, 6 Nm

3M14 A vector quantity is one
  (*a*) that has magnitude but no direction
  (*b*) whose value is known exactly
  (*c*) that has a magnitude and a direction
  (*d*) that measures quantities such as length, mass and power

3M15 If forces are equal and opposite so that no movement results, they are said to be
  (*a*) fixed   (*b*) in equilibrium
  (*c*) a resultant   (*d*) vectors

40°   40°
load 1200 N

Figure 3.11   Diagram for Exercise 3M16

3M16 A load weighs 1200 N and is suspended from a single chain, the end of which is connected to two others as shown in Figure 3.11. The tension in each of the two chains will be
(*a*) 1200 N   (*b*) 905 N   (*c*) 933 N   (*d*) 1540 N

Chapter 4

Heat

4.1 Heat

In early childhood, we become familiar with the sensations of cold and warmth, and are able to distinguish between them. We learn to estimate the degree of hotness or coldness of a body, which is known as its **temperature**.

Heat is a form of energy. Heat added to a body makes it hotter, and heat taken away from a body makes it colder. It is possible, for instance, to increase the heat energy contained in a piece of metal, and hence to increase its temperature, by doing work such as cutting, bending or hammering it. Again, if work is done against friction, or if a fire is lit beneath the metal, it will become hotter as it absorbs part of the energy made available to it. Thus, when heat energy is produced, energy in some other form is expended.

Most of the losses of energy which occur in machines appear as heat, which is usually lost to the process concerned, although not destroyed.

4.2 Temperature

Temperature is a measure of the degree of hotness or coldness of a body. We are often concerned with the accurate measurement of temperature, for which an instrument called a **thermometer** is used. Many thermometers rely for their operation on the property of many materials to expand as their temperature increases (see Section 4.6). Glass-bulb thermometers consist of a mercury- or alcohol-filled bulb attached to the bottom of a fine glass tube. The liquid rises up the tube as it expands owing to increasing temperature, which can be read off from a scale beside the tube. Increasing use is being made of thermometers which consist of a long metal strip in the form of a coil. One end is securely fixed, and movement of the other end as the strip expands moves a pointer over a scale.

Electronic thermometers have become very common, and rely on one of two operating principles. The first is that a junction between two different metals experiences a very small potential difference which depends on the temperature of the junction. This potential difference is measured by the instrument and converted to temperature which is displayed digitally. The second type uses a thermistor (a resistor whose resistance changes with temperature) enclosed in a probe which is placed at the point where the temperature is required. The probe is connected to the thermometer, which measures the resistance of the thermistor and calculates its temperature. This is displayed, usually as a digital readout. As well as being more accurate and more easily read than the other types of thermometer, the electronic type has the advantage that the protected probe can often be put in positions not accessible to normal thermometers. For example, the probe may be distant from the display, it may be driven into frozen food, it may be lowered deep into liquid, and so on.

Temperature scales

Before any scale of measurement can be decided, two fixed points are necessary. For instance, in length measurement, the distance between two marks on a metal bar can indicate a standard length. For temperature scales, the upper fixed point is the temperature of steam from boiling pure water, and the lower fixed point is the temperature at which ice just melts, both measured at normal atmospheric pressure. Between these fixed points, the scale may be subdivided into any suitable number of parts. **The Celsius or centigrade scale** takes the lower fixed point as zero degree Celsius (written 0°C) and the upper point as 100°C. **The kelvin scale** takes as its zero the lowest possible temperature (called the absolute zero), written as 0 K. A temperature change of one kelvin is the same as a change of one degree Celsius so, if we say that 0 K = – 273°C, it follows that 0°C = 273 K. Thus, any temperature expressed in degrees Celsius can be given in kelvins by the addition of 273; or any temperature given in kelvins can be converted to degrees Celsius by the subtraction of 273.

Example 4.1

(*a*) Convert to kelvins (i) 20°C (ii) 230°C (iii) – 40°C.
(*b*) Convert to degrees Celsius (i) 265 K (ii) 300 K (iii) 1200 K.

(*a*) (i) $20°C = (20 + 273)\text{kelvins} = 293 \text{ K}$
  (ii) $230°C = (230 + 273)\text{kelvins} = 503 \text{ K}$
  (iii) $– 40°C = (– 40 + 273)\text{kelvins} = 233 \text{ K}$

(*b*) (i) $265 \text{ K} = (265 – 273)\text{degrees Celsius} = – 8°C$
  (ii) $300 \text{ K} = (300 – 273)\text{degrees Celsius} = 27°C$
  (iii) $1200 \text{ K} = (1200 – 273)\text{degrees Celsius} = 927°C$

Strictly speaking, the kelvin scale is that used in SI units. However, the

Celsius scale has more convenient numbers for everyday temperatures, and its widespread use is likely to continue.

4.3   Heat units

Heat is a form of energy, so the unit of heat is the same as that for energy, the joule. It can be shown experimentally, but not proved mathematically, that to heat one litre of water (mass 1 kg) through 1°C, 4187 J of heat energy must be given to it.

   This figure 4187 is called the **specific heat** of water. Different substances have different specific heats, but, in every case, the specific heat is the heat energy in joules needed to raise the temperature of one kilogram of the substance by one kelvin. The temperature of a body depends not only on the heat contained in the body, but also on its mass and its capacity to absorb heat. For instance, a small bowl of very hot water will be at a higher temperature than a bath full of cold water, but may contain less heat energy owing to its smaller mass. Again, 1 kg of brass will increase in temperature by 1 K for the addition of 376·8 J, whereas the same mass of water requires 4187 J for this temperature change. Table 4.1 shows the specific heats of a few common materials.

Table 4.1: Specific heats

Substance	Specific heat, J/kg K	Substance	Specific heat, J/kg K
Water	4187	Copper	397
Air	1010	Brass	376
Aluminium	915	Lead	129
Iron	497		

   Thus we can see that the quantity of energy which must be added to a body to raise its temperature (or which must be removed from a body to lower its temperature) depends on its mass, its specific heat and the temperature change involved.

heat energy for temperature change, J = mass, kg × temperature change, K × specific heat, J/kg K

Example 4.2

An immersion heater is required to raise the temperature of 50 litres of water from 10°C to 85°C. If no heat is lost, find the energy required.

   energy required = mass × temperature change × specific heat

From Table 4.1, the specific heat of water is 4187 J/kg K. One litre of water has a mass of 1 kg.

$$\text{energy required} = 50 \times (85 - 10) \times 4187 \text{ J}$$

$$= 50 \times 75 \times 4187 \text{ J}$$

$$= 15\ 700\ 000 \text{ J or } 15 \cdot 7 \text{ MJ}$$

In practice, heat will be lost through the container during heating, so the efficiency of the system must be taken into account.

Example 4.3

A tank containing 150 litres of water is to be heated from 15°C to 75°C. If 25% of the heat provided is lost through the tank, how much energy must be supplied?

From Table 4.1, the specific heat of water is 4187 J/kg K. If 25% of the heat is lost, 75% is retained and the system is 75% efficient.

$$\text{energy required at } 100\% \text{ efficiency} = 150 \times (75 - 15) \times 4187 \text{ joules}$$

The energy required at 75% efficiency will be greater, and will be

$$150 \times (75 - 15) \times 4187 \times \frac{100}{75} \text{ joules}$$

$$= 150 \times 60 \times 4187 \times \frac{100}{75}$$

$$= 50\ 240\ 000 \text{ J or } 50 \cdot 24 \text{ MJ}$$

Example 4.4

2 kg of iron are placed in a furnace at an initial temperature of 15°C, and heat energy of $3 \cdot 976$ MJ is provided to the furnace. If one-quarter of the heat provided is received by the iron, what will be its final temperature?

From Table 4.1, the specific heat of iron is 497 J/kg K.

If the total heat provided is $3 \cdot 976$ MJ, and one-quarter of this reaches the iron, the heat absorbed by the iron is $3 \cdot 976/4$ megajoules, or $0 \cdot 994$ MJ

$$\text{heat provided} = \text{mass} \times \text{temperature change} \times \text{specific heat}$$

$$\text{temperature change} = \frac{\text{heat provided}}{\text{mass} \times \text{specific heat}}$$

$$= \frac{0 \cdot 994 \times 10^6}{2 \times 497} \text{ K}$$

$$= \frac{0 \cdot 994 \times 10^6}{994} \text{ K}$$

$$= 1000 \text{ K}$$

If the initial temperature is 15°C and increases by 1000 K, final temperature will be 1015°C.

4.4　Heating time and power

We have already seen that a rate of doing work of one joule per second is a power of one watt, or

$$\text{watts} \times \text{seconds} = \text{joules}$$

It is common for the kilowatt hour (kWh) to be used as a larger and more convenient unit than the joule. A kilowatt hour is the energy used when a power of 100 watts applies for one hour.

$$1 \text{ kWh} = 1000 \text{ W} \times 60 \text{ min} \times 60 \text{ s}$$

$$= 3\ 600\ 000 \text{ Ws or J}$$

or　　　　　　$1 \text{ kWh} = 3 \cdot 6 \text{ MJ}$

The rating of electric heaters is usually in watts or in kilowatts, and the time relationship of this power with heating energy enables us to calculate the time taken for a heater of a given rating to provide a given quantity of heat. Alternatively, the rating necessary to provide a given quantity of heat in a given time can be found. These calculations are illustrated in the following examples.

Although electric heating is probably the most efficient of the heating methods, some heat may be lost, and this must be taken into account.

Example 4.5

How long will a 3 kW immersion heater take to raise the temperature of 30 litres of water from 10°C to 85°C? Assume that the process is 90% efficient.

$$\text{heat required} = 30 \times (85 - 10) \times 4187 \times \frac{100}{90} \text{ joules}$$

$$= 30 \times 75 \times 4187 \times \frac{100}{90} \text{ joules}$$

$$= 10\ 470\ 000 \text{ J} \quad \text{or} \quad 10 \cdot 47 \text{ MJ}$$

A 3 kW heater gives 3000 W, or 3000 J/s.

$$\text{therefore time required} = \frac{10\ 470\ 000}{3000} \text{ seconds}$$

$$= 3490 \text{ s}$$

$$= \frac{3490}{60} \text{ minutes}$$

$$= 58 \text{ min } 10 \text{ s}$$

Example 4.6

A jointer's pot contains 25 kg of lead, and is heated by a gas flame providing a power of 2 kW to the pot. If the initial temperature of the lead is 15°C, how hot will it be after 5 min?

From Table 4.1, the specific heat of lead is 129 J/kg K. We can assume 100% efficiency, since 2 kW is provided to the pot. The heat lost will be several times this figure for a heater of this type.

$$\text{heat provided} = 2 \text{ kW for 5 min}$$

$$= 2000 \times 5 \times 60 \text{ joules}$$

$$= 600\ 000 \text{ J}$$

$$\text{heat used} = \text{mass} \times \text{temperature change} \times \text{specific heat}$$

Since heat provided = heat used,

$$600\ 000 = 25 \times \text{temperature change} \times 129$$

$$\text{temperature change} = \frac{600\ 000}{25 \times 129}$$

$$= 186 \text{ K}$$

Since the initial temperature is 15°C and the increase is 186 K, final temperature is 201°C.

Example 4.7

A 3 kW water heater is of the instantaneous type, which heats water as it flows over its element. Such a heater can be assumed to be 100% efficient. What will be the increase in water temperature from inlet to outlet if the rate of flow is 1 litre/min?

$$\text{since 3 kW} = 3000 \text{ J/s},$$

$$\text{heat provided in 1 min} = 3000 \times 60 \text{ J}$$

$$= 180\ 000 \text{ J}$$

$$\text{heat provided} = \text{mass} \times \text{temperature change} \times \text{specific heat}.$$

$$180\ 000 = 1 \times \text{temperature change} \times 4187 \text{ J}$$

$$\text{temperature change} = \frac{180\ 000}{1 \times 4187} \text{ kelvins}$$

$$= 43 \text{ K or } 43°C$$

4.5   Heat transmission

In Chapter 3, we said that some of the heat produced by a machine is 'lost' to the atmosphere. This indicates clearly that heat must be able to move from the point at which it is generated. Transmission of heat is an essential part of many engineering processes. For instance, heat must be transferred from the bit of a soldering iron to the solder and to the surfaces to be joined. Again, heat is given out from the element of an electric fire; if it were not, the continuous input of energy by means of the electric current would increase the temperature of the element until it melted.

The amount of heat transmitted depends on the difference in temperature between the heat source and the places to which the heat moves. When an electric fire is first switched on, very little heat is transmitted, because the element temperature is low. As time goes by, electrical energy is still fed into the element and appears as heat, raising the temperature and increasing the heat transmitted. In due course, the element becomes hot enough to transmit the same amount of heat energy as it receives. In this condition of **heat balance**, the temperature of the element remains constant. This heat balance applies to all devices in which heat is generated, the final steady temperature depending on the energy input and the ease with which heat can be transmitted.

In practice, heat is transmitted by three separate processes which can occur individually or in combination. These processes are called conduction, convection and radiation.

Conduction

If one end of a metal bar is heated, the other end also becomes hot. Heat has been **conducted** along the length of the bar from the high-temperature end to that at the lower temperature, the heat energy trying to distribute itself evenly throughout the bar. If we attempt a similar experiment with wood, the heated end can burn without the colder end increasing in temperature to any extent. Thus metal is said to be a good conductor of heat, whereas wood is a very poor conductor.

Metals are usually good conductors of heat, and wood, plastics and similar materials are not. For instance, a soldering iron will have a copper bit to allow maximum heat conduction to the work, but will have a wood or plastics handle to prevent it becoming too hot to hold.

Most liquids and gases are poor conductors of heat. Because of this, materials which trap air in pockets, such as felt, glass fibre and cork, are used to prevent the conduction of heat, and are called **heat insulators**. An application of such materials, often called **lagging**, is used to prevent heat escape from ovens, hot-water tanks and the like. The same sort of lagging may be used to prevent heat entering a refrigerator.

Convection

Transmission of heat by convection takes place in fluids, i.e. in liquids and gases. If a given volume of a liquid or a gas is heated, it expands if free to do so, and so the same volume weighs less than that of the unheated fluid. We can say that the density has become lower. It will tend to rise, its place being taken by cool fluid, which will also rise when heated, so that a steady upward movement of warm fluid results.

This principle is widely used in some forms of air heaters, called convectors, which draw in cool air at the bottom and expel hot air from the top.

The same sort of process is used in some types of central-heating system, where water circulates through the radiators solely as a result of convection currents. (In large installations, or those using small-bore piping, the natural circulation is assisted by a pump.)

The contents of a hot-water tank also circulate owing to convection. Water is, however, a very bad conductor of heat, and the water below the level of an immersion heater remains cold and is not displaced by convection currents. For this reason, two heaters are often fitted to hot-water cylinders. The upper heater maintains about one-third of the water hot for normal uses, the lower heater being switched on when large amounts of hot water are required.

Radiation

Most of the energy reaching the Earth does so in the form of heat radiated from the Sun. Unlike conduction and convection, which can occur only in material substances, radiation of heat from one body to another does not require any connecting medium between the bodies. Heat is radiated freely through space in much the same way as light. All bodies radiate thermal energy all the time, the rate of radiation depending mainly on the temperature of the body and its surroundings. Thus a cold body surrounded by warm objects will radiate less heat than it receives, and will tend to become warmer.

Radiant heat behaves in a very similar manner to light, and is reflected from bright surfaces. A highly polished reflector is fitted to a fire with a rod-type element, and this reflector beams the heat emitted by the element in a particular direction. Fires of this sort are often called radiators, although much of their energy is given off in the form of convection, and a little by conduction.

A body with a dark, matt surface tends to absorb heat instead of reflecting it. If the reflector of a radiator were to be painted black, it would absorb heat quickly and become excessively hot. Bright surfaces do not emit (as opposed to reflect) radiation as readily as dull ones. For this reason, appliances such as electric kettles are often chromium plated to reduce heat losses and thus improve efficiency.

4.6   Change of dimensions with temperature

Most materials increase slightly in dimensions when their temperature increases. For instance, overhead lines sag more in the summer than in the winter, and a long straight conduit run may buckle in very hot weather if expansion bends are not fitted. Similarly, dimensions often decrease when temperature decreases. An overhead cable that is erected in hot weather and strained tightly will contract in cold weather; the extra stress may then result in stretching or even breaking.

Although expansion with increasing temperature is often a nuisance, it can be applied with advantage when used to control temperature. Some metals expand more than others when heated through the same temperature range. If two strips, one of each of two metals with different rates of expansion, are riveted together, the **bimetal strip** so formed will bend when heated. If a set of contacts is operated by the strip as it bends, the device can be made to control temperature, and is called a **thermostat**. Another type of thermostat based on the same principle is called a rod-type thermostat. A rod of material, selected for its small increase in length when heated, is mounted within a tube of brass, the rod and the tube being welded together at one end. Changes in the temperature of the device result in differing changes in length of the rod and the tube, thus operating a switch. The slow break resulting from the slow rate of differential expansion will give rise to arcing at the contacts of a directly operated switch. Permanent-magnet systems and flexed springs are often used to give a quick make-and-break action to the switch (Figures 4.1 and 4.2).

Figure 4.1   Air thermostat

Figure 4.2   Rod-type immersion thermostat

4.7   Summary of formulas for Chapter 4

Temperature in kelvins = temperature in degrees Celsius + 273

Temperature in degrees Celsius = temperature in kelvins − 273

Heat energy for temperature change, J = mass, kg × temperature change, K × specific heat, J/kg K

Energy, J = power, W × time, s

4.8   Exercises

1 Express the following temperatures in kelvins:
   (a) 60°C; (b) − 75°C; (c) 1000°C.

2 Express the following temperatures in degrees Celsius:
   (a) 320 K; (b) 1500 K; (c) 240 K.

3 A small storage heater contains 8 litres of water at a temperature of 10°C. How much heat energy must be provided to raise the water temperature to 90°C? The specific heat of water is 4187 J/kg K.

4 How much heat energy must be supplied to 20 kg of brass to increase its temperature by 500 K? Give your answer in joules and in kilowatt hours. The specific heat of brass is 376 J/kg K.

5 A hall measures 10 m by 30 m, by 5 m high. How much heat energy will be required to raise the temperature of the air it contains from 5°C to 22°C? 1 m^3 of air has a mass of 1·26 kg. The specific heat of air is 1010 J/kg K. Express your answer in kilowatt hours.

6 Calculate the amount of heat in joules required to raise the temperature of 100 kg of aluminium from 10°C to 710°C. The specific heat of aluminium is 915 J/kg K.

7 40 litres of water is heated from 8°C to 78°C, the efficiency of the operation being 70%. How much heat is required? The specific heat of water is 4187 J/kg K.

8 The rate of flow of water through a water-cooled motor is 0·2 litres/s, and inlet and outlet temperatures are 10°C and 20°C, respectively. At what rate is heat being removed from the motor? The specific heat of water is 4187 J/kg K.

9 An electric-arc furnace is used to raise the temperature of 4000 kg of iron from 12°C to 812°C, the overall efficiency of the furnace being 40%. What energy input in kilowatt hours is required? The specific heat of iron is 497 J/kg K.

10 An electric water heater is 80% efficient and consumes energy at the rate of 2000 J/s. If the heater initially contains 10 litres of water at 12°C, what will be the water temperature after 10 min of heating? The specific heat of water is 4187 J/kg K.

11 The input power to a furnace for heating copper is 10 kW, and the furnace is 39·7% efficient. By how much will the temperature of 150 kg of copper have increased after 30 min? The specific heat of copper is 397 J/kg k.

12 A 2 kW heater is switched on in a room 5 m square and 3 m high, when the air temperature is 70°C. If 70% of the heat provided is lost, what will be the air temperature after 20 min? 1 m³ of air has a mass of 1·26 kg, and the specific heat of air is 1010 J/kg K.

13 How long will it take a 1·5 kW heater to raise the temperature of 10 litres of water by 60°C if the heater is 100% efficient? The specific heat of water is 4187 J/kg K.

14 An instantaneous-type water heater is required to provide a continuous flow of water of 2 litres/min at a temperature of 85°C with an inlet water temperature of 10°C. What must be the electrical loading in kilowatts if the heater is assumed to be 100% efficient? The specific heat of water is 4187 J/kg K.

15 A furnace for lead casting contains 120 kg of lead, and is heated by a 6 kW element. The initial temperature of the lead is 20°C. What temperature will the lead reach after 1 h if the furnace is 25% efficient? The specific heat of lead is 129 J/kg K.

4.9   Multiple-choice exercises

4M1 Heat is
   (a) the difference between hot and cold
   (b) temperature
   (c) a form of energy
   (d) something which is responsible for burning

4M2 Temperature is measured using
  (*a*) a thermometer
  (*b*) an ammeter
  (*c*) a sensitive human elbow
  (*d*) a thermistor

4M3 A temperature in kelvin can be converted to one in the Celsius scale by
  (*a*) subtracting 32, multiplying by 5 and dividing by 9
  (*b*) adding 273
  (*c*) using °C after the value instead of K
  (*d*) subtracting 273

4M4 The specific-heat capacity of a substance is
  (*a*) how hot it becomes when a given amount of heat is added
  (*b*) the heat energy required to raise the temperature of 1 kg of the material by 1 K
  (*c*) heat energy times mass times temperature change
  (*d*) the heat needed to cool one litre of water through 10°C

4M5 A 6 kW instantaneous heater raises the temperature of the water flowing through it from 10°C to 54°C. If the specific heat capacity of water is 4187 J/kg K, the rate of water flow in litre/min is
  (*a*) 1·59 (*b*) 1·95 (*c*) 30·7 (*d*) 0·512

4M6 Most of the energy received by the planet Earth is
  (*a*) by radiated heat from the Sun
  (*b*) a result of nuclear power stations
  (*c*) due to burning coal
  (*d*) a result of tidal action

4M7 The operation of most thermostats is a result of
  (*a*) heat convection   (*b*) electric heating
  (*c*) heat radiation   (*d*) thermal expansion

4M8 A soldering iron relies for operation on
  (*a*) soldering flux   (*b*) thermal conduction
  (*c*) operator skill   (*d*) thermal convection

4M9 An immersion heater distributes heat throughout a water storage tank as a result of
  (*a*) thermal conduction   (*b*) lagging
  (*c*) thermal convection   (*d*) thermal radiation

Chapter 5

Electrical power and energy

5.1 Units of electrical power and energy

We found in Chapter 1 that a PD of one volt exists between two points if one joule of energy is expended in moving a coulomb of electricity between them. Thus

$$\text{joules} = \text{coulombs} \times \text{volts}$$

Since the coulomb is current × time

$$\text{joules} = \text{amperes} \times \text{volts} \times \text{seconds, or } W = IUt$$

Power, as shown in Chapter 3, is the rate of using energy. The mechanical unit of power is the watt (W), and this is also the unit for electrical power. There is no reason for different units of power to exist, because the work done at a given rate can be the same whether provided by an electric motor or by a mechanical method such as a diesel engine.

The watt is defined as a rate of doing work of one joule per second.

Therefore
$$P = \frac{\text{joules}}{\text{seconds}} = \frac{IUt}{t} = IU$$

In other words, the power in watts at any instant in a given circuit is given by the instantaneous circuit current in amperes times the instantaneous circuit PD in volts, or

$$P = IU \text{ for DC circuits}$$

It is often useful to be able to express the power P in a circuit in terms of its resistance, together with either current or PD:

$$P = IU, \text{ but } U = IR, \text{ so } P = I(IR) = I^2R$$

Also
$$P = IU, \text{ but } I = \frac{U}{R}, \text{ so } P = \frac{UU}{R} = \frac{U^2}{R}$$

Note that these expressions are not necessarily true for AC circuits.

Example 5.1

A 100 Ω resistor is connected to a 10 V DC supply. What power is dissipated in it?

$$P = \frac{U^2}{R}$$

$$= \frac{10 \times 10}{100} \text{ watt}$$

$$= 1 \text{ W}$$

or

$$I = \frac{U}{R}$$

$$= \frac{10}{100} \text{ ampere}$$

$$= 0 \cdot 1 \text{ A}$$

$$P = IU$$

$$= 0 \cdot 1 \times 10 \text{ watt}$$

$$= 1 \text{ W}$$

or

$$P = I^2 R$$

$$= 0 \cdot 1 \times 0 \cdot 1 \times 100 \text{ watt}$$

$$= 1 \text{ W}$$

Example 5.2

What is the hot resistance of a 240 V, 100 W lamp?

$$P = \frac{U^2}{R}, \text{ so } R = \frac{U^2}{P}$$

$$= \frac{240 \times 240}{100} \text{ ohms}$$

$$= 576 \ \Omega$$

Since watts = joules per second, it follows that the joule, the unit of energy, is the watt second.

Example 5.3

How much energy is supplied to a 100 Ω resistor which is connected to a 150 V supply for 1 h?

$$P = \frac{U^2}{R}$$

$$= \frac{150 \times 150}{100} \text{ watts}$$

$$= 225 \text{ W}$$

$$W = Pt$$

$$= 225 \times 1 \times 60 \times 60 \text{ joules}$$

$$= 810\ 000 \text{ J}$$

Note: Time is expressed in seconds.

The joule is too small for the measurement of the amounts of energy in common use, so the **kilowatt hour** (kWh) is the unit for many practical and commercial purposes. The kilowatt hour is the energy used when a power of one kilowatt (1000 W) is used for 1 h (3600 s) and is often referred to as the 'unit' of energy.

Since joules = watts × seconds,

$$1 \text{ kWh} = 1000 \times 3600 \text{ joules} = 3\ 600\ 000 \text{ J}$$

For example 5.3, the energy is therefore 0·225 kWh.

Example 5.4

A DC motor takes 15 A from a 200 V supply, and is used for 40 min. What will this cost, if the tariff is 6 p/kWh?

$$\text{energy} = \frac{U \times I}{1000} \times \frac{\text{minutes}}{60} \text{ kilowatt hours}$$

$$= \frac{15 \times 200}{1000} \times \frac{40}{60} \text{ kilowatt hours}$$

$$= 2 \text{ kWh}$$

$$\text{energy charge} = 2 \times 6 \text{ pence} = 12\text{p}$$

Another multiple of the joule, used because the joule is small, is the megajoule (MJ). One megajoule is equal to one million joules:

$$1 \text{ MJ} = 1\ 000\ 000 \text{ J or } 10^6 \text{ J}$$

Since 1 kWh = $3 \cdot 6 \times 10^6$ joules, it follows that

$$1 \text{ kWh} = 3 \cdot 6 \text{ MJ}$$

Example 5.5

An electric heater consumes 2·7 MJ when connected to its 240 V supply for 30 min. Find the power rating of the heater, and the current taken from the supply.

$$\text{power} = \frac{\text{energy}}{\text{time}}$$

Therefore
$$P = \frac{2 \cdot 7 \times 10^6}{30 \times 60} \text{ watts}$$

$$= 1500 \text{ W or } 1 \cdot 5 \text{ kW}$$

$$P = UI \text{ so } I = \frac{P}{U}$$

$$I = \frac{1500}{240} \text{ amperes}$$

$$= 6 \cdot 25 \text{ A}$$

Example 5.6

An electric fire has two elements, each of resistance 18 Ω, and is connected to the supply through a cable of resistance 2 Ω. The supply is of 200 V. Calculate the total power taken from the supply, and the total current.

(*a*) if one element is switched on
(*b*) if both elements are switched on.

(*a*) The element resistance of 18 Ω is in series with the supply-cable resistance of 2 Ω, and can be represented as shown in Figure 5.1*a*. Since these resistances are in series, they can be represented by one 20 Ω resistance, as shown in Figure 5.1*b*.

Figure 5.1   Diagrams for Example 5.6*a*

Applying Ohm's law,

$$I = \frac{U}{R}$$

$$= \frac{200}{20} \text{ amperes}$$

$$= 10 \text{ A}$$

The power taken can be found by any of the three methods shown in Example 5.1:

$$P = UI = 200 \times 10 \text{ watts} = 2000 \text{ W}$$

or

$$P = I^2 R = 10^2 \times 20 \text{ watts} = 2000 \text{ W}$$

or

$$P = \frac{U^2}{R} = \frac{200^2}{20} \text{ watts} = 2000 \text{ W}$$

(*b*) Since the two elements in parallel are connected in series with supply cable, the circuit is effectively that shown in Figure 5.2*a*. First, find the equivalent resistance of the two 18 Ω resistors in parallel:

$$\frac{1}{R} = \frac{1}{18} + \frac{1}{18} = \frac{2}{18}$$

Therefore

$$R = \frac{18}{2} \text{ ohms} = 9 \text{ } \Omega$$

The circuit can then be redrawn as in Figure 5.2*b*, and further simplified, since the 2 Ω and 9 Ω resistors are in series, to one 11 Ω resistor connected to the 200 V supply.

From Ohm's law,

$$I = \frac{U}{R}$$

$$= \frac{200}{11} \text{ amperes}$$

$$= 18 \cdot 2 \text{ A approximately}$$

$$P = UI = 200 \times 18 \cdot 2 \text{ watts} = 3640 \text{ W}$$

Either of the methods $P = I^2 R$ or $P = U^2/R$ could be used as an alternative.

Example 5.7

How much of the total power taken from the supply by the fire of Example 5.6 is dissipated in the elements, and how much in the supply cable, if both elements are switched on?

Figure 5.2   Diagrams for Examples 5.6*b* and 5.7

If we use $P = U \times I$, the voltage used will have to be the voltage drop across the part of the circuit concerned.

Voltage drop in cable $\qquad U_c = IR$

$$= 18 \cdot 2 \times 2 \text{ volts}$$

$$= 36 \cdot 4 \text{ V}$$

Power dissipated in the cable will then be

$$P_c = U_c I = 36 \cdot 4 \times 18 \cdot 2 \text{ watts} = 662 \text{ W}$$

The power dissipated in the elements must then be the balance of the total power, so that

$$P_e = 3640 - 662 \text{ watts} = 2978 \text{ W}$$

Either of these answers could conveniently have been found by using the $P = I^2 R$ method.

$$P_c = I^2 R_c = 18 \cdot 2^2 \times 2 \text{ watts} = 662 \text{ W}$$

$$P_e = I^2 R_e = 18 \cdot 2^2 \times 9 \text{ watts} = 2978 \text{ W}$$

Example 5.8

A house installation has the following circuit loads:

lighting: two 150 W lamps, six 100 W lamps and ten 60 W lamps
ring circuit: one 3 kW and one 2 kW heaters
water heating: one 3 kW immersion heater.

What is the current in each circuit if the supply is at 240 V? What size fuse should be used for each circuit?

Lighting: Total load comprises

$$
\begin{array}{rl}
2 \times 150 \text{ watts} = & 300 \text{ W} \\
6 \times 100 \text{ watts} = & 600 \text{ W} \\
10 \times 60 \text{ watts} = & 600 \text{ W} \\
\hline
\text{total} & 1500 \text{ W}
\end{array}
$$

$$I = \frac{P}{U} = \frac{1500}{240} \text{ amperes} = 6 \cdot 25 \text{ A}$$

A suitable fuse rating would be 10 A.

Note that it is not good practice to have 18 lighting points all on the same circuit.

Ring circuit: Load comprises 2 kW + 3 kW = 5 kW

Therefore $\quad I = \dfrac{P}{U} = \dfrac{5000}{240} \text{ amperes} = 20 \cdot 8 \text{ A apporoximately.}$

A suitable fuse rating would be 30 A.

Water heating: Load comprises 3 kW

$$I = \frac{P}{U} = \frac{3000}{240} \text{ amperes} = 12 \cdot 5 \text{ A}$$

A suitable fuse rating would be 15 A.

Example 5.9

6 Ω and 3 Ω resistors are connected in parallel, this group being connected in series with a 2 Ω resistor across a DC supply. If the power dissipated in the 6 Ω resistor is 24 W, what is the supply voltage?

The circuit layout is shown in Figure 5.3, which also gives a simple method of identifying the currents and voltages.

For the 6 Ω resistor, $\qquad P_6 = 24$ W and $R_6 = 6$ Ω

$$P_6 = I_6^2 R_6, \text{ so } I_6^2 = \frac{P_6}{R_6} = \frac{24}{6} \text{ A}^2 = 4 \text{ A}^2$$

Therefore $\qquad\qquad I_6 = \sqrt{I_6^2} = \sqrt{4} \text{ A} = 2 \text{ A}$

Thus $\qquad\qquad U_6 = I_6 R_6 = 2 \times 6 \text{ volts} = 12 \text{ V}$

Since the 6 Ω and 3 Ω resistors are in parallel,

$$U_6 = U_3 = 12 \text{ V}$$

Therefore $\qquad\qquad I_3 = \frac{U_3}{R_3} = \frac{12}{3} \text{ amperes} = 4 \text{ A}$

But I_6 and I_3 are made up from current I_2, which splits after passing through R_2.

Thus $\qquad\qquad I_2 = I_6 + I_3 = 2 + 4 \text{ amperes} = 6 \text{ A}$

Therefore $\qquad\qquad U_2 = I_2 R_2 = 6 \times 2 \text{ volts} = 12 \text{ V}$

Figure 5.3 Diagram for Example 5.9

Supply voltage V is made up of V_2 in series with the PD across the parallel bank, V_6 or V_3.

Therefore $\qquad\qquad\qquad U = U_2 + U_6 = 12 + 12 \text{ volts} = 24 \text{ V}$

5.2   Electromechanical conversions

In SI units, there is no difference between mechanical and electrical quantities. Thus the watt is the unit of power in all systems, and is always defined as a rate of doing work of one joule per second. Similarly, the joule is the unit of energy in both mechanical and electrical systems. In mechanical applications, the joule is defined as the work that is done when a force of one newton is moved through one metre. For electrical applications, the joule can be defined as the work that is done when one coulomb of electricity is moved through a potential difference of one volt.

Since the same units are used for both systems, conversions from mechanical to electrical forms, and vice versa, are simple, relying on the following relationships:

$$\text{joules} = \text{coulombs} \times \text{volts} = \text{metres} \times \text{newtons}$$

Since coulombs = amperes × seconds,

$$\text{watts} = \frac{\text{joules}}{\text{seconds}} = \text{amperes} \times \text{volts} = \frac{\text{metres} \times \text{newtons}}{\text{seconds}}$$

Example 5.10

Calculate the power rating of a lift which can raise a mass of 800 kg through a height of 50 m in $98 \cdot 1$ s.

$$\text{work done} = \text{force (newtons)} \times \text{distance (metres)}$$

1 kg requires a force of $9 \cdot 81$ N to lift it against gravity, so 800 kg require $800 \times 9 \cdot 81$ N.

$$\text{work done} = 800 \times 9 \cdot 81 \times 50 \text{ J}$$

$$\text{power (W)} = \frac{\text{work done, J}}{\text{time, s}}$$

$$= \frac{800 \times 9 \cdot 81 \times 50}{98 \cdot 1} \text{ watts}$$

$$= 4000 \text{ W or } 4 \text{ kW}$$

Example 5.11

Calculate the output and input powers of a motor driving the lift of Example 5.10 if the lift gear is 80% efficient and the motor is 90% efficient.

$$efficiency = \frac{output}{input}$$

Input to the lift (which is also the output of the motor) $= \frac{output}{efficiency}$

$$output\ of\ motor = 4 \times \frac{100}{80}\ kilowatts = 5\ kW$$

$$motor\ input = \frac{output}{efficiency}$$

$$= 5 \times \frac{100}{90}\ kilowatts$$

$$= 5 \cdot 56\ kW$$

Example 5.12

If the motor of Examples 5.10 and 5.11 is fed from a 220 V DC supply, what current will it take during the lift?

$$P = UI\ so\ I = \frac{P}{U}$$

$$I = \frac{5560}{220}\ amperes = 25 \cdot 2\ A$$

Example 5.13

A DC generator is to provide a maximum current of 500 A at 220 V and is $0 \cdot 88$ efficient. What must be the kilowatt rating of the diesel engine driving it?

$$output = I \times U = 500 \times 220\ W$$

$$input = \frac{output}{efficiency}$$

$$= \frac{500 \times 220}{0 \cdot 88}\ watts$$

$$= \frac{500 \times 220}{0 \cdot 88 \times 1000}\ kilowatts$$

$$= 125\ kW$$

5.3   Electric lighting and heating

We have already seen that, when an electric current passes through a resistor, power is produced. This power appears as heat in the resistor.

The production of heat due to dissipation of electrical energy is widespread, and examples are numerous. A few of the most common are listed below:

Filament lamps

A very thin tungsten wire is formed into a small coil and supported within a glass bulb. The passage of an electric current through this filament causes it to reach a temperature of 2500°C or more, so that it glows brightly. At these temperatures, the oxygen in the atmosphere would combine with the filament to cause failure, so all the air is removed from the glass bulb and replaced by a gas such as nitrogen, which does not react with the hot tungsten. The construction of a filament lamp is shown in Figure 5.4.

gas filling
(argon and nitrogen)

glass bulb

tungsten filament

molybdenum
supports

lead wire

glass pinch

exhaust tube

fuse sleeve
and fuse

cement

exhaust-tube seal

cap

soldered contact

Figure 5.4   Tungsten-filament lamp

Air heaters

The purpose of these devices is to warm the air in buildings, and thus make living and working more comfortable in cold weather. As shown in Chapter 4, the heat is transferred to the air from the resistive element by radiation and convection.

The radiant heater has an element that operates at a temperature high enough for it to glow red. Heat is radiated, and is often directed to the required area by the use of a polished reflector.

A convector heater usually has a low-temperature 'black-heat' element. Air in contact with the element is warmed and becomes less dense, so that it rises and is replaced by colder air, which is warmed in turn. A continuous output of warm air results.

A forced-convection, or fan-type, heater is shown in Figure 5.5.

Owing to the reduced cost of electricity taken only at times of low demand, storage heaters are widely used. The radiator type is heated by elements in fireclay blocks. The heat from these blocks is conducted through the lagging provided, which slows down the rate at which heat is released. The fan-type storage heater has thicker lagging, so that little heat is lost through it. Heat is provided by passing air over the heated blocks by means of a fan. Such a heater is shown in Figure 5.6. The solid floor of a building may also be used to store heat, heating cables being buried in it.

Figure 5.5   Fan-type heater

Figure 5.6   Cut-away view of fan-type storage heater

Water heaters

An electric element is enclosed in a copper or nickel tube, and electrically insulated from it with a mineral insulation of magnesium oxide. The resulting heater is immersed in the water to be heated. Heat is transferred to the water by conduction, after which the water circulates naturally by a process of convection. There are many types of electric water heater, one of the most common being shown in Figure 5.7.

Other heating applications

There are very many more applications of electric heating in industry, in commerce and at home. For example, electric welding is widely used for fabrication, and electric furnaces are used to produce steel and other alloys.

In the home, the electric cooker, kettle, iron and radiator are electric heaters that are often taken for granted.

Figure 5.7   Nonpressure or free-outlet-type water heater

5.4   Summary of formulas for Chapter 5

$$P = \frac{W,}{t} = IU = I^2R = \frac{U^2}{R}$$

where

P = power dissipated, W
W = energy expended, J
t = time taken, s
I = circuit current, A
R = circuit resistance, Ω
U = circuit potential difference, V

Note: $P = IU$ and $P = U^2/R$ are not always true for AC circuits.

$$1 \text{ kWh} = 3\ 600\ 000 \text{ J} = 3 \cdot 6 \text{ MJ}$$

5.5   Exercises

1 An electric fire takes a current of $6 \cdot 25$ A from a 240 V supply. What power does it dissipate?

2 How much power will be dissipated in a 3 Ω resistor when it is connected across a 12 V battery?

3 A coil of resistance 5 Ω carries a current of 10 A. What power is dissipated in the coil?

4 A hall is lit by 20 lamps, each of which takes $0 \cdot 5$ A when connected to a 240 V supply. Calculate the total power supply to the lamps.

5 What is the resistance of an electric-fire element rated at 1 kW when carrying a current of 5 A?

6 When it is dissipating a power of 1 W, the PD across a carbon radio resistor is 100 V. What is the resistance?

7 What current will a 750 W fire element take from a 240 V supply?

8 How much current will a 10 Ω resistor be carrying if the power lost in the resistor is 1 kW?

9 A 3 kW heating element has a resistance of 30 Ω. For what supply voltage was the heater designed?

10 A 10 kΩ resistor in an electronic circuit has $0 \cdot 5$ W of power lost in it. What current is it carrying, and what is the PD across it?

11 How much energy will be consumed by the resistor of Exercise 10 in 10 min?

12 An electric water heater must provide 4 MJ of energy to heat its contents in 50 min. What is the power rating of the heating element?

13 An underfloor heating system has a loading of 20 kW. How much energy will it consume in 10 h?

14 A 10 Ω resistor carries a current of 5 A. How much energy will it dissipate as heat each minute?

15 A 2 Ω resistor liberates as heat 1440 J of energy in 20 s. What is the PD across the resistor?

16 A motor with an output of 8 kW and an efficiency of 80% runs for 8 h. Calculate the cost if electricity costs 6p per unit.

17 A generator in a motor car charges the battery at 20 A at 16 V. The generator efficiency is 43%. What power is absorbed in driving it?

18 The rating of the hotplate of an electric cooker is unknown. At 6 p.m. the meter reads 14 567 units. The cooker hotplate is at once switched on, together

with a 1 kW fire and 500 W of lighting, and left on until 10 p.m., when the meter reads 14 581 units. What is the rating of the hotplate?

19 It costs 40p to run a motor for 2 h, the tariff being 6·67p per unit. If the motor is 90% efficient, what is its output?

20 An electric water heater provides 48 MJ in 40 min. What is the rating of the element, and how much current does it take from a 240 V supply?

21 A workshop supplied at 250 V has two motor drives at 1 kW each, a fan which takes 1 A, two 100 W lamps, a 2 kW radiator and a battery charger which takes 50 W. Calculate the total load in kilowatts and the weekly cost at 6p per unit if the load is on for 40 h a week.

22 A resistor of unknown value is connected in series with two 4 Ω resistors, which are connected in parallel, across a 30 V supply. The power dissipated in each 4 Ω resistor is 16 W. Calculate the value of the unknown resistor.

23 Resistors of 5 Ω, 10 Ω and 30 Ω are connected in parallel. The combination is connected in series with a 7 Ω resistor across a 50 V supply. Calculate the power dissipated in each resistor.

24 A 60 Ω resistor and a resistor of unknown value are connected in series with each other and with a parallel combination of a 200 Ω and a 50 Ω resistor. This circuit is connected to a 240 V supply. If the power dissipated in the circuit is 480 W, what is the value of the unknown resistor?

25 An electric radiator has a resistance of 19 Ω and is connected to a supply by a cable of resistance 1 Ω. A second radiator of 38 Ω resistance is connected to the supply by a cable of resistance 2 Ω. The supply voltage is 240 V. Calculate
(*a*) the current in each radiator,
(*b*) the total power taken from the mains.

26 A domestic load comprises twelve 60 W lamps, eight 100 W lamps, two 2 kW radiators and a 3 kW immersion heater. The lamps are fed from a lighting circuit and the power from a ring-main circuit. The supply is at 240 V. What is the current in each circuit? Suggest a suitable fuse for each circuit from ratings of 2 A, 5 A, 10 A, 13 A, 20 A and 30 A.

27 A small flat has the following loads:
3 × 100 W lamps in use 5 h each day
1 × 2 kW heater in use 4 h each day
1 × 3 kW immersion heater in use 3 h each day
1 × 4 kW cooker in use 2 h each day.
The tariff for the supply of electricity is a fixed charge of 18p per week, plus an energy charge of 5p per unit. Find the total electricity cost for one week.

28 A water heater is 83·7% efficient, and is required to heat 12 litres of water from 15°C to 75°C in 1 h. What minimum rating is required for the element? The specific heat of water is 4187 J/kg K.

29 How long will a 3 kW heater take to raise the temperature of 30 litres of water from 20°C to 90°C if 10% of the heat supplied is lost in the process? The specific heat of water is 4187 J/kg K.

30 A builder's hoist lifts 100 kg of bricks through 40 m in 20 s. If the hoist is 75% efficient, what is the rating of the electric motor driving it?

31 If the motor of the previous exercise is fed from a 200 V DC supply and is 80% efficient, calculate its running current.

32 A lift motor takes a current of 20 A from a 300 V DC supply. The lift gearing is 50% efficient. If the lift is able to raise a load of 1440 N through a height of 50 m in 30 s, what is the efficiency of the motor?

33 Calculate the output of a motor with an input of 24 kW and an efficiency of 90%.

34 A motor with an output of 10 kW and an efficiency of 85% drives a drilling machine which is 60% efficient. What is the output of the machine, and what is the input of the motor?

5.6   Multiple-choice exercises

5M1 The unit of electrical power is the
(a) volt (b) ampere (c) watt (d) horsepower

5M2 The power dissipated in an 800 Ω resistor connected to a 240 V supply is
(a) 72 W (b) 0·3 A (c) 7·2 W (d) 192 W

5M3 An electric lamp is rated at 60 W. The current it draws from a 240 V supply is
(a) 4 A (b) 14·4 mA (c) 6·94 μA (d) 0·25 A

5M4 If the 60 W lamp of question 5M3 is left on for eight hours and electrical energy costs 7p per unit, the cost will be
(a) 3·36p (b) £3·36 (c) 5·25p (d) 6·86p

5M5 The energy used by the 60 W lamp of Exercise 5M4 in eight hours is
(a) 480 J (b) 172·8 kJ (c) 28·8 kJ (d) 60 W

5M6 A resistor of value 33 Ω is connected to a 100 V supply through a cable with a total resistance of 0·3 Ω. The thermal energy in the cable is
(a) 300 W (b) 3·63 W (c) 2·71 W (d) 33·3 kW

5M7 Resistors are connected in a series–parallel group as shown in the figure below. The power dissipated in the 5 Ω resistor is
(a) 2·56 W (b) 45 W (c) 80 W (d) 4 A

5M8 The SI unit of force is the
(*a*) pound (*b*) horsepower (*c*) joule (*d*) newton

5M9 The number of newtons in one kilogram is
(*a*) 9·81 (*b*) 33 000 (*c*) 453·6 (*d*) 3 600 000

5M10 If 120 J of energy is used to move a force of 10 N, the distance moved is
(*a*) 1·2 m (*b*) 12 m (*c*) 120 cm (*d*) 0·083 m

5M11 A block and tackle system is 40% efficient and lifts a load of 300 kg to
a height of 4 m in 10 s. The input power required is
(*a*) 3 kW (*b*) 4·70 kW (*c*) 2·94 kW (*d*) 480 W

5M12 The operation of a filament lamp depends on
(*a*) the fact that certain powders emit light (fluoresce) when stimulated
(*b*) the filament being heated to a temperature which causes it to glow and
to emit visible light
(*c*) a gas jet being lighted
(*d*) a bayonet-cap lampholder

5M13 The water in an electrical water heater is warmed from the element by
the process of
(*a*) convection   (*b*) radiation
(*c*) a copper or nickel tube   (*d*) conduction

5M14 A fan-type thermal-storage heater is less expensive to run than a direct
electric heater because
(*a*) it has a fan which circulates air through it
(*b*) it is usually provided with input and fan controls
(*c*) it is charged with cheaper electrical energy at offpeak times
(*d*) it has very effective thermal insulation

Chapter 6

Permanent magnetism and electromagnetism

6.1 Magnetic fields

Every schoolboy is familiar with permanent magnets, which can be used to pick up small iron objects. Such a magnet gives rise to a magnetic field which occupies the space in which the effects of the magnet can be detected. The magnetic field extends outwards in the space surrounding the magnet, getting weaker as the distance from the magnet increases. There are several methods of detecting the presence of a magnetic field, although it is quite invisible and does not affect other human senses. Theoretically, the magnetic field due to a magnet extends for considerable distances, but, in practice, it will combine with the fields of other magnets to form a composite field, so that the effects of most magnets can only be detected quite close to them.

Because we cannot see, feel, smell, or hear it, a magnetic field is difficult to represent. A 'picture' of a magnetic field is often useful, however, in deciding what its effects will be, and to do this we consider that a magnetic field is made up of imaginary **lines of magnetic flux** which have the following properties:

(*a*) They always form complete, closed loops.
(*b*) They never cross one another.
(*c*) They have a definite direction.
(*d*) They try to contract as if they were stretched elastic threads.
(*e*) They repel one another when lying side by side and having the same direction.

It is most important to remember that any consideration involving lines of magnetic flux is quite imaginary. The lines of magnetic flux do not exist, but to pretend that they do gives us a method of understanding the behaviour of a magnetic field.

Experiments with crude permanent magnets were carried out many

109

hundreds of years ago by the Chinese. They found that, if a magnet is suspended so that it can pivot freely, it will come to rest with a particular part (one end, if it is a bar magnet) pointing north. This is the **north-seeking pole** or **north pole** of the magnet, and the other end, which points south, is the **south-seeking pole** or **south pole**. If two magnets, with their poles marked, are brought together, the basic laws of magnetic attraction and repulsion can easily be demonstrated. These are:

(i) *Like poles repel:* For instance, two north poles or two south poles will try to push apart from each other.

(ii) *Unlike poles attract:* For instance, a north pole and a south pole will attract one another.

The Earth behaves as if a huge bar magnet were inside it, the south pole of the magnet being near the geographic North Pole, and the north pole at the geographic South Pole. The positions of the magnetic north and south poles of the Earth vary slowly with time, for reasons not yet fully explained. As far as we can tell, however, the Earth is a truly permanent magnet. If a piece of soft iron is magnetised by rubbing it in one direction with one pole of a permanent magnet, the soft iron will lose its magnetism slowly as time goes by. The magnetism could be removed quickly by heating or hammering the soft iron. Magnets which do not deteriorate with time, and which resist demagnetisation by ill treatment, are now available, and are mentioned later in this chapter.

The pattern of the lines of magnetic flux around a permanent bar magnet are shown in Figure 6.1. This shows that the direction of the lines of magnetic flux (property *c*) is from north pole to south pole outside the magnet. These lines of magnetic flux obey the five rules given, their curved shapes outside the magnet being a compromise between properties *d* and *e*.

Figure 6.2 shows the flux pattern due to two magnets with unlike poles close together. Lines of magnetic flux are imagined to try to contract (property *d*), and the magnets try to pull together.

Figure 6.1   Magnetic field due to bar magnet

Figure 6.2   Magnetic field due to two bar magnets with unlike poles adjacent

Figure 6.3 shows the magnetic-flux pattern due to two magnets with like poles close together. Since magnetic flux lines running side by side with the same direction repel, the two poles try to push apart.

These explanations indicate the neat way in which the results of magnetic fields can be forecast by the use of lines of magnetic flux. These lines can be plotted for an actual magnetic field in two ways. The first is to cover the magnet system concerned with a sheet of paper and then to sprinkle on iron filings, which set themselves in the magnetic-flux pattern. The second method involves the use of a plotting compass, which consists of a miniature magnetic needle pivoted within a clear container. The compass is placed on the paper, and the position of its north and south poles marked with a pencil. It is then moved, and its position adjusted until its south pole is at the mark indicating the previous position of the north pole. The new position of the north pole is marked, and the process repeated as often as is necessary. The line of marks which results is joined to form a line of magnetic flux. As many lines as are required can be produced in a similar way.

Figure 6.3   Magnetic field due to two bar magnets with like poles adjacent

6.2   Units of magnetic flux

If the imaginary lines of magnetic flux already introduced actually existed, they could be counted and used as a measure of the quantity of magnetism. They do not exist, but the idea of a larger number of lines representing a greater magnetic flux than a smaller number is a useful one.

The general symbol for magnetic flux is Φ (Greek letter, capital 'phi'). The unit of magnetic flux is the **weber** (pronounced 'vayber' and abbreviated to Wb). It must be clearly understood that the weber is a measure of a total amount, or quantity, of magnetic flux, and not a measure of its concentration or density. Flux density is very important in some machines, and will depend on the amount of magnetic flux (the number of lines, in effect) which is concentrated in a given cross-sectional area of the flux path. The strength of a magnetic field is measured in terms of its **flux density** (symbol B), measured in **webers per square metre** (Wb/m^2), also called **teslas** (T). Thus, one weber of magnetic flux spread evenly throughout a cross-sectional area of one square metre results in a flux density of one tesla, or one weber per square metre. Similarly 1 Wb spread over 10 m^2 will give a flux density of 0·1 T.

Thus
$$B = \frac{\Phi}{A}$$

where B = magnetic-flux density, T (Wb/m^2)
  Φ = magnetic flux, Wb
  A = cross-sectional area of flux path, m^2

It should be appreciated that magnetic-flux path areas of the order of square metres are only met in the largest machines, but small quantities of flux in small areas can give rise to high magnetic flux densities.

Example 6.1

The magnetic flux per pole in a DC machine is 2 mWb, and the effective poleface dimensions are 0·1 m × 0·2 m. Find the average flux density at the poleface.

Since the same submultiple prefixes apply to magnetic flux as to other quantities, 2 mWb = 0·002 Wb.

$$\text{effective poleface area} = 0·1 \times 0·2 \text{ square metres}$$

$$= 0·02 \text{ m}^2$$

$$B = \frac{\Phi}{A}$$

$$= \frac{0·002}{0·02} \text{ tesla}$$

$$= 0·1 \text{ T}$$

Example 6.2

A lifting electromagnet has a working magnetic flux density of 1 T, and the effective area of one poleface is a circle of diameter 70 mm. What is the total magnetic flux produced?

effective poleface area $= \dfrac{\pi d^2}{4}$

$$= \pi \times \dfrac{70 \times 70}{4} \text{ square millimetres}$$

$$= 3850 \text{ mm}^2$$

$$= \dfrac{3850}{1000 \times 1000} \text{ square metres}$$

$$= 0\cdot00385 \text{ m}^2$$

$$\Phi = BA$$

$$= 1 \times 0\cdot00385 \text{ webers}$$

$$= 0\cdot00385 \text{ Wb}$$

$$= 3\cdot85 \text{ mWb}$$

6.3   Electromagnet

If an electric conductor is arranged to pass vertically through a horizontal sheet of paper, iron filings or a plotting compass will show a field pattern at the paper if a sufficiently large current flows through the conductor. Since both filings and compass are affected only by a magnetic field, this indicates that the current in the conductor must be producing the field; this observation is borne out in practice, because the effect on both filings and compass disappears when the current is switched off.

The arrangement described is shown in Figure 6.4. This shows only a small part of the magnetic field, however; the field extends for the full length of the conductor in the form of concentric rings of magnetic flux centred on the conductor. It is often necessary to indicate the direction of current flow in a cross-section of a cable, since cable and resultant magnetic field form a three-dimensional arrangement. The convention for such an

direction of current flow

Figure 6.4   Plotting magnetic field due to current-carrying conductor

indication is shown in Figure 6.5, current flowing away from the viewer (into the paper) being shown by a cross, and current flowing towards the viewer (out of the paper) being shown as a dot. This can easily be remembered if the current is replaced by a dart sliding in a hollow tube which represents the conductor. When the dart is sliding away from the viewer, its flights are seen as a cross, and, when sliding towards the viewer, the point is seen as a dot.

Using a plotting compass, the magnetic fields round current-carrying conductors can be shown to have the directions given in Figure 6.5; that is, clockwise for current flowing away from, and anticlockwise for current flowing towards, the viewer. This can easily be remembered by the **screw rule** which states that, if a normal right-hand-thread screw is driven along the conductor in the direction taken by the current, its direction of rotation will be the direction of the magnetic field. It is quite common in electrical work for current-carrying conductors to lie side by side in a cable or conduit. Figure 6.6 shows that the magnetic field in the space between the conductors may be quite intense if they carry current in opposite directions, although the magnetic field in an enclosing steel conduit will be in different directions in opposite sides of the conduit. If one or more conductors in a conduit carry current in the same direction, most of the magnetic flux will be restricted to the conduit, and may give rise to heating losses. This is the reason why feed and return cables should be enclosed in the same conduit.

The strength of the magnetic field round a conductor depends on the current, but, even at high currents, it is comparatively weak. To obtain

Figure 6.5 Direction of magnetic field round current-carrying conductor

Figure 6.6 Magnetic field due to two current-carrying conductors

a stronger field, the magnetic effects of a number of conductors can be added. The most common form for this arrangement is one long insulated conductor would in a tight coil called a **solenoid**. The form taken by the solenoid is shown in Figure 6.7a; Figure 6.7b is a cross-section of the solenoid showing how the individual magnetic fields due to each separate turn merge to form a stronger field that is very similar to that of the permanent bar magnet (Figure 6.1). The strength of the magnetic field produced depends on the current and on the number of turns used. Additional turns may be wound on top of the first layer to form a multilayer coil.

The polarity of the solenoid can be found easily by sketching it with correct current directions (Figure 6.7b), or by using the **NS rule**. If arrows drawn on the ends of a capital N point in the direction of current flow in the solenoid when viewed from one end, that end is a north pole. If arrows on a capital S give the direction of current, the solenoid end viewed is a south pole (Figure 6.8).

The solenoid is the electrical equivalent of the permanent magnet, but is often more versatile. For instance, the flux and flux density can be altered by a variation of solenoid current, or reduced to zero by switching off the current. Since the solenoid is simply a coil of conductor, its size and shape can be constructed to almost any requirements. The electromagnet, based on the solenoid, is the basis of many items of electrical equipment, of which the relay, the contactor, the motor, the generator, the transformer and the telephone are examples.

Figure 6.7   Solenoid and its magnetic field

Figure 6.8   NS rule

6.4   Calculations for aircored solenoids

If we multiply the number of turns of a solenoid by the current it carries, we arrive at the **magnetomotive force** (MMF) of the solenoid, which is measured in **ampere turns** (At) or, more strictly, in amperes (A):

$$\text{magnetomotive force} = \text{amperes} \times \text{turns}$$

For instance, two of the methods of producing an MMF of, say, 1000 At are 1 A flowing in a 1000-turn coil, or 10 A flowing in a 100-turn coil.

The magnetomotive force obviously affects the flux set up by the solenoid, but so, too, does the length of path taken by individual lines of magnetic flux. For instance, a 1000-turn coil wound in four layers and having a length of, say, $0 \cdot 1$ m will set up more flux for a given current than a 1000-turn coil wound in one layer over a length of $0 \cdot 4$ m. The path taken by the lines of magnetic flux will be longer in the second case, so that the magnetomotive force is 'stretched' over a greater distance. The magnetomotive force applied to each metre length of the path taken by the lines of magnetic flux is called the **magnetising force** (symbol H), measured in **ampere turns per metre** (At/m) or, more strictly, in amperes per metre (A/m)

$$H = \frac{\text{amperes} \times \text{turns}}{\text{length of flux path in metres}}$$

or

$$H = \frac{IN}{l} \text{ ampere turns per metre}$$

where H = magnetising force, At/m or A/m
$\quad\quad I$ = solenoid current, A
$\quad\quad N$ = number of turns of conductor on solenoid
$\quad\quad l$ = mean length of magnetic-flux path, m

The path taken by the lines of magnetic flux is often referred to as the **magnetic circuit**.

Example 6.3

A solenoid, wound with wire having an overall diameter of 1 mm, must have 12 layers of winding, each 100 mm long. Calculate how much current must flow in the coil to give a magnetising force of 3000 At/m in a magnetic circuit of average length $0 \cdot 25$ m which has the coil mounted on it.

$$\text{number of turns per layer} = \frac{\text{solenoid length}}{\text{wire diameter}} = \frac{100 \text{ mm}}{1 \text{ mm}} = 100 \text{ turns}$$

Total turns = 1200

$$H = \frac{IN}{l}$$

Therefore
$$I = \frac{Hl}{N}$$

$$= \frac{300 \times 0 \cdot 25}{1200} \text{ amperes}$$

$$= 0 \cdot 625 \text{ A}$$

When a current flows in a coil, the resulting magnetising force sets up a magnetic flux. The amount of flux set up in air and other materials which are not attracted by a magnet will depend directly on the applied magnetising force. For instance, the extremely high magnetising force of one million ampere turns per metre would produce within the solenoid a flux density of $4\pi/10$ or $1 \cdot 257$ T. Thus the ratio

$$\frac{B}{H} = \frac{4\pi/10}{10^6} = \frac{4\pi}{10^7} \text{ or } 4\pi \times 10^{-7}$$

This constant relationship between flux density and magnetising force for air is called the **permeability of free space**, and is given the symbol μ_0 (μ is the Greek letter 'mu').

$$\mu_0 = \frac{4\pi}{10^7}$$

Example 6.4

An aircored solenoid is in the form of a closed ring (or toroid) of mean length $0 \cdot 2$ m and cross-sectional area 1000 mm^2. It is wound with 1000 turns and carries a current of 2 A. Find the magnetic-flux density and total magnetic flux produced within the solenoid.

$$H = \frac{IN}{l} = \frac{2 \times 1000}{0 \cdot 2} \text{ ampere turns per metre} = 10\,000 \text{ At/m}$$

$$\frac{B}{H} = \mu_0$$

$$B = \mu_0 H$$

$$= \frac{4\pi}{10^7} \times 10\,000 \text{ tesla}$$

$$= 0 \cdot 01257 \text{ T}$$

or $12 \cdot 57$ mT

total flux $\Phi = BA$

$$= 0 \cdot 01257 \times \frac{1000}{10^6} \text{ weber } (\textit{Note:} \ 1 \text{ m}^2 = 10^6 \text{ mm}^2)$$

$$= 0 \cdot 00001257 \text{ Wb}$$

or $12 \cdot 57$ μWb

Notice the low magnetic-flux density and total flux set up by the comparatively high magnetising force.

Example 6.5

A solenoid of 10 000 turns is wound on a brass ring of mean diameter 100 mm and cross-sectional area $0\cdot01$ m^2. How much current must flow in the solenoid to produce a total flux of 5 mWb?

Brass is a nonmagnetic material, so the rules given will apply

$$B = \frac{\Phi}{A}$$

$$= \frac{0\cdot005}{0\cdot01} \text{ tesla}$$

$$= 0\cdot5 \text{ T}$$

$$\frac{B}{H} = \mu_0$$

so

$$H = \frac{B}{\mu_0}$$

$$= \frac{IN}{l}$$

$$H = \frac{0\cdot5 \times 10^7}{4\pi} \text{ ampere turns per metre}$$

Also

$$H = \frac{I \times 10\ 000}{\pi/10} \text{ ampere turns per metre}$$

$$I = \frac{0\cdot5 \times 10^7 \times \pi}{4\pi \times 10\ 000 \times 10} \text{ amperes}$$

$$= 12\cdot5 \text{ A}$$

6.5   Effect of iron on magnetic circuit

Most materials in use for general purposes are nonmagnetic; that is, they show no magnetic properties. A simple test for magnetic properties is to see if a material is attracted to a permanent magnet, or to an electromagnet. The materials which show this attraction are iron, nickel and cobalt, as well as alloys containing one or all of them.

If a core of magnetic material is slid into a solenoid, the magnetic flux produced for the same current will be increased very many times. The ratio of flux produced by a solenoid with a magnetic core to flux produced by the same solenoid with an aircore (the current being the same in both cases) is called the **relative permeability** (symbol μ_r) of the material under these conditions. The value of the relative permeability of a nonmagnetic material is unity (one), whereas values for magnetic materials vary between wide limits, typical figures for common magnetic materials being from about 150 to about 1200. This value is not a constant for a given material, since it depends on the magnetising force applied. Taking relative permeability into account,

$$\frac{B}{H} = \mu_0\mu_r$$

The product $\mu_0\mu_r$ is called the **absolute permeability** of the material under the given conditions.

In many applications, a magnetic circuit has a small airgap in it. In some, such as moving-coil instruments or machines, this is because free movement must be possible between different parts of the magnetic circuit. In other machines, poorly fitting joints in the magnetic circuit have the same effect as an airgap. Even a small airgap may have a disproportionate effect on a magnetic circuit. Example 6.6 illustrates the advantage of using a magnetic material for a magnetic circuit, and the reduction of flux that occurs if an airgap is introduced.

Example 6.6

(a) A solenoid is made to be identical with that described in Example 6.4, but is wound on a wrought-iron core. When the solenoid current is 0·2 A, relative permeability of the wrought iron is 500. Calculate the flux density and total flux in the ring.

(b) What would be the effect on flux and flux density if a radial sawcut were to be made in the wrought-iron core without disturbing the magnetising coil?

(a) $$H = \frac{IN}{l}$$

$$= \frac{0\cdot2 \times 1000}{0\cdot2} \text{ ampere turns per metre}$$

$$= 1000 \text{ At/m}$$

$$\frac{B}{H} = \mu_0\mu_r$$

so $$B = \mu_0\mu_r H$$

Therefore $$B = \frac{4\pi}{10^7} \times 500 \times 1000 \text{ tesla}$$

$$= 0\cdot628 \text{ T}$$

$$\Phi = BA$$

$$= 0 \cdot 628 \times 0 \cdot 001 \text{ weber}$$

$$= 0 \cdot 000628 \text{ weber}$$

$$= 0 \cdot 628 \text{ mWb}$$

(*b*) The introduction of an airgap into the wrought-iron magnetic circuit will reduce considerably the magnetic flux and flux density set up by the magnetising coil. Calculations of flux values in a case of this sort are complicated, but it can be shown that, if the sawcut is only 3 mm wide, flux density will be reduced to $0 \cdot 0714$ T and flux to $0 \cdot 0714$ mWb.

The results of this example are worthy of close examination, since they show some of the principles of magnetic-circuit design. Part *a* shows that the substitution of a magnetic material as the core of the solenoid has increased the flux and flux density by 50 times, even though the magnetising current is one-tenth its previous value.

The answer to part *b* shows that the airgap (length 3 mm) makes the establishment of magnetic flux nearly ten times as difficult as when the iron path was complete. This is despite the fact that $98 \cdot 5\%$ of the circuit (197 mm) still consists of iron.

6.6   Permanent magnets

Permanent-magnet materials have been the subject of intese development over the past few years, and magnets can now be made that really are permanent, despite the most adverse conditions of temperature, mechanical ill treatment and demagnetising fields. The materials used are the three main magnetic elements (iron, nickel and cobalt), with the addition of other elements such as tungsten, carbon, manganese, chromium, copper, aluminium etc. Modern permanent magnets are often too hard to drill or machine, and must be cast in their final correct shapes. Intricate magnetic circuits often use a small permanent magnet at the centre of soft-iron extension pieces.

Although modern permanent magnets are unlikely to change their properties during normal use, they can sometimes be damaged magnetically when removed from their surrounding magnetic circuit. If it is necessary to remove a permanent magnet, the following precautions should be observed:

(*a*) Complete the magnetic circuit using a mild-steel 'keeper'.
(*b*) Do not slide the keeper against the magnet; remove it with a direct pull.
(*c*) Do not touch the magnet with steel tools, such as screwdrivers or spanners.
(*d*) Do not put the magnet on a steel-topped workbench.

(e) Keep the magnet in a nonmagnetic tray or box, well separated from other magnets.

(f) Do not allow people to play with the magnet.

6.7   Summary of formulas for Chapter 6

$$B = \frac{\Phi}{A} \qquad \Phi = BA \qquad A = \frac{\Phi}{B}$$

where B = magnetic-flux density, T (Wb/m^2)
  Φ = magnetic flux, Wb
  A = cross-sectional area of flux path, m^2

$$H = \frac{IN}{l} \qquad I = \frac{Hl}{N} \qquad N = \frac{Hl}{I}$$

where H = magnetising force (ampere turns per metre, or amperes per metre)
  I = magnetising current, A
  N = number of magnetising coil turns
  l = mean length of magnetic-flux path, m

Permeability of free space

$$\mu_0 = \frac{4\pi}{10^7}$$

$$\mu_0 = \frac{B}{H} \qquad B = \mu_0 H \qquad H = \frac{B}{\mu_0}$$

For magnetic materials

$$\mu_0 \mu_r = \frac{B}{H} \qquad B = \mu_0 \mu_r H \qquad H = \frac{B}{\mu_0 \mu_r}$$

where μ_r is the relative permeability of the magnetic material under the given conditions.

6.8   Exercises

1 It is required to determine the magnetic field surrounding a bar magnet. Give a brief description of a laboratory procedure to do this.

2 Draw a diagram to show a solenoid wound over an iron core. Mark on the diagram the magnetic lines of flux resulting from current in the solenoid. Show a direction for the current, and give the resulting direction of the flux lines.

3 Draw circles to represent the cross-sections of two conductors lying side by side. Mark the directions of currents in the conductors, and sketch the resulting magnetic field if the two conductors carry currents
(*a*) in the same direction
(*b*) in the opposite directions.

4 The total flux required in the core of a power transformer is $0 \cdot 156$ Wb. If the area of the core is $0 \cdot 12$ m^2, what will be the flux density?

5 A moving-coil instrument has an airgap of effective cross-sectional area 56 mm^2. What is the total flux if the gap flux density is $0 \cdot 12$ T?

6 A coil of 400 turns wound over a wooden ring of mean circumference $0 \cdot 5$ m and uniform cross-sectional area 800 mm^2 carries a current of 6 A. Calculate
(*a*) magnetising force
(*b*) total flux
(*c*) flux density.

7 If an airgap of length 20 mm is introduced into a magnetic circuit of cross-sectional area $0 \cdot 009$ m^2 that is carrying a flux of $1 \cdot 2$ mWb, what extra magnetomotive force will be required to maintain the flux?

8 Calculate the flux and flux density set up by the solenoid of exercise 6 if the wooden ring is replaced by one of magnetic material having identical dimensions and a relative permeability of 200 at this flux density.

9 A coil of insulated wire of 500 turns and of resistance 4 Ω is closely wound on an iron ring. The ring has a mean diameter of $0 \cdot 25$ m and a uniform cross-sectional area of 700 mm^2. Calculate the total flux in the ring when a DC supply at 6 V is applied to the ends of the winding. Assume a relative permeability of 550. Explain the general effect of making a small airgap by cutting the iron ring radially at one point.

10 Make sketches indicating current direction and resulting magnetic field when current flows in
(*a*) a straight conductor
(*b*) a solenoid about 40 mm in diameter and 50 mm long
(*c*) a similar solenoid with a 20 mm diameter steel core 50 mm long
(*d*) a solenoid with a steel core in the shape of a closed ring.

11 Figure 6.9 shows a coil of wire wound on to an iron core with an airgap in one limb. How would the magnetic flux in the gap be affected by an increase in
(*a*) the current in the coil
(*b*) the number of turns of wire
(*c*) the length of the iron path
(*d*) the length of the gap
(*e*) the permeability of the iron?

6.9 Multiple-choice exercises

6M1 A space in which magnetic effects can be detected is called
(*a*) permeability   (*b*) a magnetic field
(*c*) magnetic flux   (*d*) lines of force

6M2 Lines of magnetic flux are considered to have a number of properties. The property in the following list which is INCORRECT is that
(*a*) they have a definite direction
(*b*) they always form complete closed loops
(*c*) they never cross one another
(*d*) they attract each other when lying side by side and having the same direction

6M3 If the south-seeking poles of two magnets are brought close together, they will
(*a*) repel each other   (*b*) be unaffected
(*c*) pull together   (*d*) attract each other

6M4 Lines of magnetic flux are considered to have the direction
(*a*) towards a north pole
(*b*) north to south outside a magnet
(*c*) towards a south pole
(*d*) south to north outside a magnet

6M5 Magnetised soft iron may lose its magnetism if
(*a*) immersed in water
(*b*) placed in a drawer
(*c*) heated or hammered
(*d*) bent

6M6 The shape of a magnetic field can be found by using
(*a*) a plotting compass or iron filings
(*b*) an Ordnance Survey map
(*c*) a land surveyor
(*d*) an ammeter

6M7 The symbol used to represent magnetic flux density is
(*a*) T   (*b*) *D*   (*c*) Wb   (*d*) *B*

6M8 If a flux density of $0 \cdot 16$ T appears at the 20×10 cm poleface of a machine, the total flux is
(*a*) 32 Wb   (*b*) $3 \cdot 2$ mWb   (*c*) 8 T   (*d*) $312 \cdot 5$ mWb

6M9 A current flowing towards the viewer in a cross-sectioned conductor is shown by
(*a*) clockwise concentric circles
(*b*) a cross
(*c*) a dot
(*d*) anticlockwise concentric circles

Figure 6.9   Diagram for Exercise 11

6M10 If two adjacent conductors are carrying current in the same direction, they will both experience a force which
(a) pushes them apart
(b) tries to move them lengthways
(c) heats them up
(d) pulls them together

6M11 A solenoid is
(a) a kind of heat lamp for browning the skin
(b) a type of permanent magnet
(c) a coil of wire which sets up a magnetic field when it carries an electric current
(d) part of a boot

6M12 The formula for magnetising force is

(a) $H = \dfrac{IN}{l}$ (b) $\dfrac{B}{H} = \mu_0\mu_r$ (c) $B = \dfrac{H}{A}$ (d) $\mu_0 = 4\pi \times 10^{-7}$

6M13 A magnetic circuit has a mean length of $0 \cdot 4$ m, a cross-sectional area of 120 mm^2 and is made of copper. If wound with 2500 turns of wire carrying a current of 120 mA, the total flux set up in the copper will be
(a) 750 A/m   (b) $0 \cdot 113$ µWb   (c) $7 \cdot 85$ Wb   (d) $0 \cdot 036$ Wb

6M14 The formula relating absolute permeability, relative permeability, magnetic flux density and magnetising force is

(a) $H = \dfrac{IN}{l}$ (b) $\mu_0\mu_r = BH$ (c) $A = \dfrac{H}{B}$ (d) $\dfrac{B}{H} = \mu_0\mu_r$

Chapter 7

Applications of electromagnetism

7.1 Introduction

This chapter will describe some of the applications of electromagnetism which are commonly in use. It should be noted, however, that no complete list of such applications is possible, because this list is almost endless. The principles put forward in Chapter 6 apply to all these devices.

Two of the most important of all the electromagnetic equipments are not mentioned in this chapter. These are the generator and the motor, which are introduced in Chapters 9 and 11, respectively.

7.2 Bells and buzzers

The operation of bells and buzzers is identical, and units of these types are used widely to give audible warnings. To a large extent, electromagnetic bells and buzzers have now been replaced by electronic sounders, but they still hold a very important place. There are a number of types of electromagnetic bell.

The **single-stroke bell** has applications where simple signalling is necessary. It consists of an electromagnet which, when energised, attracts a soft-iron strip, called an armature, against the pull of a flat spring strip. A striker supported by the armature sounds the gong (Figure 7.1). When the current is switched off by releasing the signalling push, the spring strip returns the armature to its original position.

The **trembler bell** is the most widely used type, and is similar in construction to the single-stroke bell, but has the addition of a set of contacts P which are opened by movement of the armature towards the electromagnet (Figure 7.2). This de-energises the magnet, and allows the armature to return to its original position, closing the contacts and repeating the cycle. While the supply is maintained, the armature is in a continual state of vibration, its striker hitting the gong repeatedly. The rate of striking depends

Figure 7.1 Single-stroke bell

Figure 7.2 Trembler bell

on the flexibility of the spring strip, and the weight and size of the moving system.

The **buzzer** has no striker and normally oscillates more rapidly; some adjustment of movement, and hence of buzzer note, being possible with the contact screw.

The **continuous-ringing bell** is particularly useful for alarm systems, since it will continue to ring after the external operating circuit has been broken. The construction (Figure 7.3) includes a relay arm which drops when the bell first rings, closing an internal operating circuit, which continues to ring the bell until reset by lifting the relay arm to its original position. This can be done by means of a pull cord or by a solenoid, operated remotely.

Figure 7.3   Continuous-ringing bell

The **polarised bell** is mainly used for telephone circuits where a low-frequency AC supply is used. A permanent-magnet system (Figure 7.4) is fitted with two operating coils connected in series but wound in opposite directions. Alternating current will alternately weaken and strengthen the magnetic fields in the side limbs, one being strengthened while the other is weakened. The moving system will thus be attracted to each side in turn, the striker sounding each bell in turn.

7.3   Bell indicators and circuits

There are many applications of systems which require that a large number of widely spaced pushes should be used to indicate, at one central position,

Figure 7.4   Polarised bell

that service is required. Hospitals, hotels, restaurants and many other semipublic buildings need to be so equipped. Fire-alarm and burglar-alarm systems are other examples of installations where the point of origin of the alarm may need to be indicated at some central or remote point.

The range of bells and buzzers with differing tones is strictly limited, so the indicator board was developed to cater for this situation. Usually, only one bell or buzzer sounds, the board giving a clear indication of the position of the push that was operated. Electronic indicator boards, with filament lamps or light emitting diodes (LEDs) which are kept on by electronic latching circuits until reset, are very common. These are not electromagnetic devices and thus have no place in this chapter. Electromagnetic types are still common, however, so their operation is described.

There are three basic types of electromagnetic indicator unit:

(a) **Pendulum-type indicator:** The indicator 'flag' is attracted to a solenoid while the press is pushed, falling away and swinging to give indication when it is released. This is a simple and inexpensive system, but the indication ceases after a time when the flag stops swinging. The pendulum type is largely replaced by the other two types.

(b) **Mechanical-reset-type indicator:** The coloured disc of this unit is not normally seen through the window on the front of the indicator, but is pulled into view by a solenoid coil when the circuit concerned is energised. It then remains in the indicating position until reset by the mechanical operation of a rod or lever.

(c) **Electrical-reset-type indicator:** This is similar in operation to the mechanical-reset type, but is returned to the normal position by a reset solenoid operated from a control push normally situated adjacent to the board (Figure 7.5).

Units of these types are made up into boards containing any required number of ways, each way suitably labelled.

Figure 7.5   Electrical reset-type bell-indicator unit

Figure 7.6 shows how a four-way indicator board can be connected. Additional boards may be connected in series with the first if indication at other points is necessary.

Figure 7.6   Four-way indicator unit and circuit

7.4   Relays and contactors

Relays

A relay is similar in construction to the single-stroke bell, but a set of contacts are opened or closed by operation, instead of a gong being struck. This enables one circuit to operate another (Figure 7.7), a very low current often being sufficient to close the contacts of a relay, and thus operate a high-power circuit. Some relays have a large number of contacts, and can be used in complicated circuits for a wide variety of switching purposes. As in other applications, electronics is becoming common in electromagnetics, and many relays are now solid state, using thyristors or triacs.

Figure 7.7   Principle of relays and contactors

A relay usually has an operating coil wound on a magnetic circuit with a moving section, or armature, held open by a spring. When the operating current is switched on and sets up a magnetic field, the armature is closed and operates the contacts.

Contactors

Contactors are simply very large relays, enabling a heavy load to be switched on and off with a very small operating current. A few applications are

(*a*) motor starting: the contactor coil, and hence the motor, is controlled by 'stop' and 'start' pushbuttons; the contactor also ensures that the supply leads to the motor are broken in the event of a supply failure, so that the motor cannot restart automatically when the supply is restored

(*b*) timeswitch contactor: a heavy heating load can be switched by a timeswitch with low-rated contacts; the timeswitch controls the coil of the contactor, which controls the load

(*c*) remote switching: the principle is shown in Figure 7.7.

7.5   Telephones

Operation of a telephone is based on the fact that noise consists of a series of pressure waves in the air. These pressure changes are caused by the vocal chords in speech, and alternate bands of high and low pressure move outwards in ever widening circles from the noise source, like the ripples on a pond into which a stone has been thrown. These waves vibrate the eardrums, making it possible to hear the sound.

The telephone changes these variations in pressure into corresponding changes in an electric current, which can be made to flow for long distances before giving up energy to convert the current variations back into pressure changes.

Microphone or transmitter

This is the device used to change sound into an electric current. Its operation depends on the fact that a mass of small **carbon granules** has electrical resistance which varies with the sound pressure applied to it.

The construction, shown in simplified form in Figure 7.8, consists of two flat discs of solid carbon, one fixed solidly at the back of the microphone capsule and one fixed to the back of a thin metal diaphragm (usually of aluminium). The diaphragm vibrates in sympathy with the pressure waves striking it, and thus the carbon granules packed between the two carbon discs are subject to pressure variations. If a battery is connected as shown, the current will vary with the resistance of the granule pack, its waveshape being very similar to that of the sound concerned. For ease of servicing, microphone capsules are normally self-contained units which plug into the telephone handset. Figure 7.9 shows in diagrammatic form the type of transmitter used in most public telephones.

Figure 7.8   Simplified diagram of carbon-granule microphone capsule

Figure 7.9   Typical telephone transmitter inset

Receiver

This is the device which converts the varying electric current back into sound waves, and consists basically of a permanent magnet wound with coils of fine wire (Figure 7.10). A thin diaphragm of magnetic material is held in position by the permanent magnet, very close to its poles but not touching them. When the current from the microphone passes through the coils, the magnetic field varies in sympathy with it, so that the diaphragm is vibrated. This vibration sets up pressure waves in the air, corresponding closely to those at the microphone.

Figure 7.10   Simplified diagram of permanent-magnet receiver

Circuit

If two microphones and two receivers were connected in series with a battery (Figure 7.11), a conversation could be held. However, the useful distance over which the system would operate satisfactorily would be very limited. There is no ringing system, and the batteries would soon be exhausted, since they are in circuit all the time. Some simple circuits which overcome these difficulties are given in Section 7.6.

Figure 7.11   Elementary telephone circuit

7.6   Simple telephone circuits

A typical simple circuit for two telephones is shown in Figure 7.12. If the press key of one telephone is operated, the circuit is completed as shown by the open arrows, the battery associated with one telephone ringing the bell of the other. It should be noted that both handsets must be in position for the ringing circuit to be complete, since lifting either set results in a break in the circuit. When both sets are lifted, however, the speaking circuit is completed as shown by the closed arrows.

The power loss in a resistive line is proportional to the square of the current flowing, so that, if the current ·an be reduced ten times, the loss will be reduced 100 times. A longer circuit is possible here, the reduction of current being obtained by using induction coils. A complete circuit of this type is shown in Figure 7.13, from which it should be noted that the receiver is connected in the low-current system, where it operates quite

Figure 7.12   Simple two-instrument telephone circuit

Figure 7.13   Simple telephone circuit with induction coils

satisfactorily. The induction coils are, in fact, simple transformers, which convert the high varying direct current of the microphone circuit to a low-current, higher-voltage system of the same waveshape in the remainder of the circuit.

It must be stressed that the circuits given have been considerably simplified to show the basic principles concerned. Practical circuits can be very complex indeed, with large numbers of telephones in communication with each other and with so-called 'master' sets, which often have special facilities for priority and secret conversations. Telephone circuits today are largely electronic systems which are beyond the scope of this volume. However, the circuit principles are based on those explained here.

The use of loudspeaker-type units is increasing, although the field is inclined to be limited by the publicity involved. A loudspeaker-type receiver and a particularly sensitive microphone take the place of the normal handset in such units.

7.7   Loudspeakers

The loudspeaker is used to convert electrical signals into sound waves. The very weak electrical pulses received by a radio aerial, for instance, are amplified in the electronic circuits of the radio set, before being fed to the loudspeaker so that the programme can be heard.

A typical loudspeaker has a magnetic circuit comprising a permanent magnet, and completed by soft-iron pole pieces, so that a strong magnetic flux exists in a short cylindrical airgap. A moving coil, called the speech or voice coil, is suspended from the end of a paper or plastics cone so that it lies in the gap (Figure 7.14). Alternating currents in the coil push it up and down, vibrating the cone and producing sound waves.

The excellence of a loudspeaker depends on the strength of the magnetic field in the gap, the lightness of the coil, which is often wound of aluminium wire, and the size of the cone. A large cone produces low-frequency notes well, and a small cone deals better with high-frequency notes. Some loudspeakers have two cones to enable them to reproduce notes faithfully over a wide range of frequencies.

Figure 7.14   Simple construction of permanent-magnet moving-coil loudspeaker

7.8   Moving-iron instruments

The moving-iron instrument is a device used for the measurement of electrical current, which can also be adapted for use as a voltmeter. There are two basic types of moving-iron instrument.

Moving-iron repulsion instrument

In its simplest form, this instrument consists of a magnetising coil which carries the current to be measured, inside which are two parallel soft-iron bars (Figure 7.15). One of the bars is fixed to the coil, and the other is attached to the moving system of the instrument and is free to swing with it. When current flows in the coil, both bars are magnetised to an extent which depends on the current. Since the bars are magnetised by the same field, their magnetic polarities will be the same, with both north poles together at one end and both south poles together at the other. The repulsion which results moves the bar which is free to swing, and gives a deflection.

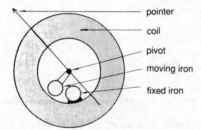

pointer
coil
pivot
moving iron
fixed iron

Figure 7.15   Principle of moving-iron repulsion instrument

Moving-iron attraction instrument

The magnetising coil of this instrument is wound so that the central space will accept a soft-iron vane which is attracted to it (Figure 7.16). The vane is attached to the moving system, and thus indicates the strength of the magnetic field, and hence of the current producing it.

pointer
coil
pivot
vane

Figure 7.16   Principle of moving-iron attraction instrument

All moving-iron instruments have the major advantage of reading on both DC and AC supplies. A coil fed with an alternating current results in an alternating magnetic field, which continues to attract the soft-iron vane, or to reverse the direction of the magnetisation of the two soft-iron bars together, so that they continue to repel. These instruments are robust and comparatively cheap, but have a serious disadvantage, in that their scales follow a 'square law' and are very cramped at the lower end. Little can be done to improve the lower end of the scale, but the rest of it can be made more linear by shaping the disc of the attraction type, or substituting shaped plates for soft-iron bars in the repulsion type. Other disadvantages are that moving-iron instruments are not as accurate as permanent-magnet moving-coil types (Section 11.6), and are affected by stray magnetic fields and by changes in the frequency of an AC supply.

7.9   Exercises

1 Describe the construction and operation of
   (*a*) an electric bell
   (*b*) an electric buzzer.

2 Describe the construction and operation of a single-stroke bell; indicate one application of such a bell.

3 Using a carefully drawn sketch, show the construction of a continuous-ringing bell. Explain how it functions, and where it could be used.

4 Describe one type of bell indicator. Draw a circuit diagram for a four-way indicator system, using one bell, an indicator board and four pushes.

5 Explain the construction and operation of a relay. How can this device allow heavy currents to be switched remotely, using small conductors to the remote position?

6 Suggest four applications for a relay or contactor, and explain each.

7 Describe, with the aid of a sketch, the construction and operation of a simple electromagnetic relay suitable for a bell circuit.

8 Make a neat, labelled sketch to show the construction and operation of *one* of the following:
   (*a*) a telephone receiver (earpiece)
   (*b*) a buzzer.

9 With the aid of a sketch, give a brief description of the construction of a telephone receiver.

10 Describe with the aid of a sketch the construction and operation of a telephone microphone.

11 Draw a circuit showing two telephones connected together. Each telephone must have provision for calling the other, as well as the speaking circuit. Explain how the circuit functions.

12 Using a clearly labelled diagram, describe the construction and operation of a loudspeaker.

13 Describe the construction and operation of one type of moving-iron instrument. What are its advantages and disadvantages when compared with the permanent-magnet moving-coil instrument?

7.10 Multiple-choice exercises

7M1 The trembler bell relies for its operation on
 (*a*) the application of thyristors and triacs
 (*b*) an electromagnetic resetting system
 (*c*) contacts which open on operation, allowing the system to repeat its action
 (*d*) the presence of a permanent magnet to give polarisation

7M2 The buzzer differs from the bell in construction because it
 (*a*) makes a different noise
 (*b*) has no bell or striker and thus oscillates more rapidly
 (*c*) is very much smaller
 (*d*) is of the electrical-reset type

7M3 A continuous-ringing bell is one which
 (*a*) is designed to ring forever
 (*b*) cannot be stopped once it has started to ring
 (*c*) depends for its operation on magnetic repulsion
 (*d*) once started, continues to ring until reset

7M4 A bell or buzzer indicator system is used to
 (*a*) show from which position the bell or buzzer was operated
 (*b*) make an installation more complicated
 (*c*) handle current levels too high for the bell or buzzer alone
 (*d*) allow more bells or buzzers to be installed than is strictly necessary

7M5 A polarised bell is usually used for
 (*a*) a house door bell   (*b*) very cold situations
 (*c*) a telephone   (*d*) a permanent magnet

7M6 A relay can be useful to
 (*a*) promote an athletics meeting
 (*b*) allow a set of heavy-current contacts to be closed by a small current
 (*c*) make a telephone easier to hear
 (*d*) change an old cable for a new one

7M7 The microphone or transmitter of a telephone relies for its operation on
 (*a*) the British Telecom or Mercury networks
 (*b*) a change in the resistance of carbon granules when subjected to a change in pressure
 (*c*) the changing magnetic field produced by an electromagnet
 (*d*) shouting loudly into it

7M8   The receiver of a telephone depends for its operation on
    (*a*) the changing magnetic field produced by coils wound on a permanent magnet
    (*b*) a change in resistance of carbon granules when subject to a change in pressure
    (*c*) the British Telecom or Mercury networks
    (*d*) the surrounding sound level being very low

7M9   A loudspeaker with a very small cone is best at reproducing
    (*a*) pop music
    (*b*) very low-frequency notes
    (*c*) classical music
    (*d*) high-frequency notes

7M10   The moving-iron repulsion instrument operates because
    (*a*) the current to be measured repels the moving system
    (*b*) two fixed irons are oppositely magnetised
    (*c*) a fixed and a moving iron are magnetised with the same polarity
    (*d*) a soft-iron vane is attracted into a shaped coil

7M11   The moving-iron attraction instrument operates because
    (*a*) a soft-iron vane is attracted into a shaped coil
    (*b*) it has a needle moving over a scale
    (*c*) two fixed irons are oppositely magnetised
    (*d*) the scale is nonlinear

7M12   A moving-iron instrument will operate when the supply is
    (*a*) alternating current only
    (*b*) direct current only
    (*c*) alternating or direct current
    (*d*) very large

Chapter 8

Electric cells and batteries

8.1 Storing electricity

Most of the electrical power used is generated in rotating machines; a sufficient number of generators are necessary to provide the maximum load required, since power used at a particular time must be generated at that time. Many generators stand idle for long periods because they are needed only to meet the peak demands, which usually occur for only a few hours a day. Fewer generators would be necessary if electricity could be stored during the night for use during the day. In one method, a pumped-storage system is used, in which water is pumped up to a high-level reservoir during the period of low demand at night so that it can be used to drive water turbines at times when demand is high. Lack of suitable sites prevents the system from being widely applied. Another method of storage is to provide chemical energy which can be converted to electrical energy as required. Although the cost of such an operation on a national scale would be prohibitive, this method is very widely used. A unit for chemical to electrical energy conversion is called a **cell**, and there are two types of cell.

Primary cells have the active chemicals placed in them when they are made. When the chemicals are used up, they are sometimes replaced, but it is more usual to throw away the spent cell. Most primary cells are now made in the dry form, and are widely used in torches, portable radio sets and the like.

Secondary cells are capable of being reactivated when their energy is spent. This is done by passing a charging current of electricity through them. Some of this energy is converted to the chemical form, and stored for future use. Secondary cells and batteries are widely used to power small, portable equipments on vehicles, for standby supplies in case of mains failure, and for similar purposes.

8.2   Primary cells

Primary cells are those which have their energy added in chemical form during manufacture, and which normally cannot be recharged once this energy is spent. The easiest way to understand primary cells is to take first the simplest of them and to examine its defects, showing how these are overcome in more complex cells.

Simple cell

Figure 8.1 shows a simple cell, which consists of plates of copper and zinc immersed in a weak solution of sulphuric acid. If an external circuit is connected across the plates, current flows from the copper (positive) plate to the zinc (negative) plate, the circuit being completed through the sulphuric acid, which is called the **electrolyte**. Hydrogen is produced in this action and collects on the copper plate in the form of fine bubbles, which effectively insulate the plate from the electrolyte. This effect is called **polarisation**, and it results in a sharp decrease in cell EMF from its initial value of $1 \cdot 08$ V, when a current is drawn.

Figure 8.1   Simple electric cell

The zinc plate is often in a form which is far from pure; small particles of other metals, such as iron and lead, are embedded in it. When the plate is immersed in the electrolyte, these impurities, in conjunction with the zinc, form tiny cells on its surface, and the zinc plate is eroded away. This effect is called **local action**. These two disadvantages make the simple cell unsuitable for practical use. Practical cells use pure-zinc plates, or a coating of mercury on the zinc, to prevent local action. Polarisation is overcome by placing the positive electrode in a chemical which absorbs the hydrogen, and is called a **depolariser**.

Wet Leclanché cell

Cells of this type are still in operation in some bell and telephone circuits. A section of a typical cell is shown in Figure 8.2. The mercury-coated zinc

Figure 8.2   Leclanché cell (wet)

(negative) rod is enclosed within the glass jar, in its electrolyte of ammonium chloride, sometimes called sal ammoniac. The carbon (positive) rod is packed in a depolariser of crushed carbon and manganese dioxide, separated from the electrolyte by a porous pot. The depolariser is efficient, but rather slow in action, so that the cell is most useful for intermittent operation. The electrolyte has a tendency to creep up the sides of the glass jar. This is prevented by painting or greasing the inside of the upper part of the jar. The EMF of this cell is 1·5 V.

Zinc-chloride cell

This is a form of Leclanché cell in which the liquid is replaced by a paste, and which can thus be used in any position. The sheet-zinc container, often slid into a cardboard tube, holds the electrolyte, which consists of plaster of paris saturated with ammonium chloride (Figure 8.3). Since the

Figure 8.3   Zinc-chloride cell

electrolyte is in paste form, it does not mix with the crushed carbon and manganese-dioxide depolariser, from which it is separated by enclosing the latter in a linen bag. At the centre of the bag is the carbon (positive) rod, which is usually surmounted by a brass cap to make easy contact. The cell is sealed with pitch, a vent being left for the escape of gases. The cell EMF is $1 \cdot 5$ V when new, falling to a steady $1 \cdot 4$ V in service. When the cell voltage falls to 1 V, it should be discarded, since, at this stage, the zinc container quickly corrodes and allows its contents to escape. **Leakproof** cells enclosed in a second steel case are available to prevent this nuisance.

Alkaline cells

The alkaline manganese cell has a voltage of $1 \cdot 5$ V and, although more expensive, has a number of useful advantages, including:

1 self venting to prevent bursting in the event of a sustained short-circuit
2 long shelf life
3 a much more uniform voltage than the zinc-chloride types
4 very good leakage protection
5 low internal resistance
6 can provide comparatively heavy current for long periods without the need for a rest period to recover
7 will operate satisfactorily over a very wide range of ambient temperatures (typically $-20°C$ to $+70°C$).

The construction of a typical alkaline cell is shown in Figure 8.4.

Mercury cells

These $1 \cdot 35$ V cells have the advantage of very small size, often being made in the form of 'button cells'. However, the toxic nature of mercury means that their use is declining.

Lithium cells

With a range of voltages from $1 \cdot 5$ V to $3 \cdot 7$ V, a single lithium cell is often able to replace two or more cells of other types. They have exceptionally long shelf and operating lives, sometimes in excess of ten years, so they are used in equipments where the frequent need to change batteries would be a problem, such as cameras and portable computing devices. Their ability to store energy is almost three times that of other cells of the same size, they have a much wider operating temperature span and they have a much greater ability to provide high currents. However, the extremely high cost of the cells limits wider use.

cathode cap

insulating washer

outer steel jacket

separator

anode of powdered zinc

electrolyte of potassium hydroxide

cathode of mixed manganese dioxide and graphite

cathode collector

anode collector

plastic grommet

vent

insulator

anode cap

Figure 8.4　Alkaline cell

8.3　Secondary cells

There are three basic types of secondary cell: lead–acid, alkaline and those small types which are used to power transportable electronic equipments such as telephones, calculators, lap-top computers and so on. These types have a reversible chemical action and can be recharged by passing an electric current through them.

The characteristics, maintenance and charging of secondary cells is covered in Section 8.4.

Lead–acid cells

Lead–acid cells are the most widely used type of secondary cell, and consist of two plates of lead in an electrolyte of dilute sulphuric acid (flooded type). The positive plate is often in the form of a lead–antimony-alloy lattice into which active lead oxide is pressed. The negative plate is usually of pure lead. Water is produced on discharge, and lowers the **specific gravity** of the dilute sulphuric acid. Specific gravity is a method of indicating the strength of the electrolyte, and is the ratio of the mass of a given volume of the electrolyte to the mass of the same volume of water. Measurement of specific gravity using a **hydrometer** (see Figure 8.7) is a good method of determining the state of charge. If the cell is overdischarged, or allowed to stand for long periods in the discharged condition, the lead-sulphate coating on the plates becomes hard and is difficult to remove by charging. In this condition, the efficiency of the cell is reduced, and it is said to be **sulphated**. A healthy lead–acid cell has an initial EMF in the region of $2 \cdot 2$ V, which falls to 2 V when in use.

The construction of lead–acid cells varies with the duty they are required to perform. Cells for use in vehicles are usually made up into batteries, a number of cells being included in separate containers in the same moulding of hard rubber, celluloid or polystyrene. Terminals for each cell are brought through the top of the moulding, which includes removable vent plugs for inspection and topping up. To obtain maximum capacity for minimum volume, the plates are mounted close together but prevented from touching by separators of thin wood or porous plastics.

Stationary cells are permanently installed in buildings to provide power for systems which must still operate when the mains supply fails, including telephones, emergency lighting, bells, alarm systems and so on. Since stationary cells are unlikely to be subjected to mechanical shocks, the containers are usually in the form of glass or plastics containers, with active plates and separators suspended within them. Containers are usually mounted on glass or porcelain insulators to limit leakage currents.

All lead–acid cells have space left below the plates for accumulation of active material forced out of the plates while in use, and which might otherwise 'short' out the cell. The specific gravity of a charged cell varies with its type, and is also slightly affected by temperature, but an average value is $1 \cdot 250$. All lead–acid batteries pose dangers to the maintainer, and these are considered in Section 8.4.

Conventional lead–acid cells emit oxygen and hydrogen during operation, the gases being replaced by 'topping up' with water (H_2O). A recent development has been the 'maintenance-free' cell, containing recombination plugs to convert the hydrogen and oxygen to water inside the cell. This allows the electrolyte to be bonded in a thixotropic gel rather than a liquid. It is necessary for the cell to be provided with a valve to allow escape of excess hydrogen which results from short-circuit conditions, and to prevent the entry of oxygen which would result in internal corrosion. Cells and batteries of this type require little maintenance and are used to power small transportable equipments, as car batteries, for standby purposes and so on.

Nickel–cadmium cells

These cells are enclosed in plastics or steel cases and use potassium hydroxide as an electrolyte. The cell has interleaved flat plates, positive of nickel

filler caps

vent plug:
designed to eliminate spray but
give free exit of gases

cell pillars and connectors:
each one designed specifically
for the job; give minimum
resistance—maximum current
flow

cell lids:
opaque SAN

bar guard:
safeguards against
short circuits

negative plates:
pasted grids; provide perfect
balance with the positive to give
maximum performance

separators:
sintered microporous PVC
gives minimum resistance

Planté positive plates:
constructed of pure lead to
ensure that there is no fall off in
capacity throughout their long
life

plastic containers:
transparent SAN:
electrolyte level and cell
condition clearly seen

Figure 8.5   Flooded-type lead–acid stationary cell

hydroxide and negative of cadmium. Figure 8.6 shows the construction of this type of cell. The specific gravity of the electrolyte remains steady during discharge at about $1 \cdot 20$ but may fall with age; the electrolyte should be replaced when its specific gravity falls to $1 \cdot 16$.

The robustness of nickel–cadmium cells and their resistance to both mechanical and electrical ill treatment make them suitable in the form of batteries for some types of electric vehicle, military use, the operation of high-voltage switchgear, railway signalling and all purposes where failure cannot be accepted. These cells will maintain their charge for very long periods without attention, and can operate over a range of temperatures which would make other types ineffective. High initial cost, wide

terminal pillar — vent cap

— filler cap

cell cover — steel gland nut
(welded joint)

— stuffing box

— gland ring

gas-release valve —

assembly bolt — — spacing washer
— locked nut

positive-
plate frame — negative-
plate frame

positive plate — positive
active material

negative
active material

negative plate —

steel container —
(specially rustproofed) — insulating rods

— edge insulator

finely perforated — suspension boss
pocket envelopes

cell bottom
(welded joint)

Figure 8.6   Construction of nickel–cadmium cell

terminal-voltage range (from $1 \cdot 4$ V charged to $1 \cdot 1$ V discharged), higher internal resistance (more than double that of the comparable lead–acid battery), make them less popular for general use than lead–acid types.

Small sealed nickel–cadmium rechargeable cells are being used increasingly in both their normal and button configurations because they are small, light and extremely reliable. Some types have the ability to be recharged very rapidly (within minutes). They have very long life, are maintenance free, may be overcharged without ill effect, can be stored for long periods in either charged or discharged condition, have a high discharge rate and constant discharge voltage and are mechanically very robust. These advantages make them the usual choice for applications such as cordless tools, camcorders, mobile telephones, mobile computers and so on.

8.4   Care of secondary cells

When the chemicals of a cell have changed to the inactive form, they can be made active once more by passing a charging current through the battery in the opposite direction to the discharge current. The supply voltage must be in excess of the battery voltage, or no charging current can flow. There are two methods of charging:

(*a*) *Constant-voltage charging:* A constant voltage is applied to the battery under charge. While the battery is charging, a steady increase occurs in its terminal voltage, so that the charging current tapers off to a lower value at the end of the charge than at the beginning. This system has the advantage that no adjustments are required during charging; however, since it is recommended that lead–acid batteries are better charged at a constant rate, this method is seldom employed in the UK, although it is widely used in Continental Europe.

(*b*) *Constant-current charging:* This system uses either an adjustable voltage source or a variable resistance, so that the charging current can be kept constant throughout the charge. Healthier stationary batteries result from this method, but the variation concerned normally requires manual adjustment, although automatic means can be provided. A wide range of specialist computer-controlled chargers is available from battery manufacturers which monitor the battery condition and adjust charging current to provide the optimum charge and to ensure that there will be no overcharging. It should be appreciated that cells and batteries are very temperature conscious, and that correct charging levels will change if there is a change in the battery temperature.

Capacity

The total charge which a battery or a cell will hold is measured in terms of the current which is supplied, multiplied by the time for which it flows. In practice, this figure varies depending on the rate of discharge, a quick discharge giving a lower figure than a slow one. Capacity (which should not be confused with capacitance) is usually measured at the ten-hour rate; for instance, a sixty ampere-hour (60 Ah) battery will provide six amperes for ten hours.

Charging and maintenance of lead–acid and alkaline cells are quite different, and will be considered separately.

Lead–acid cells: maintenance

These cells are the most widely used of the secondary cells, owing to their comparatively low cost and higher voltage per cell. They can be damaged, however, by charging or discharging too quickly, overcharging, leaving in the discharged state etc. Healthy cells can only be maintained in condition either by keeping them fully charged, or by periodically recharging, ideally at monthly intervals. A lead–acid cell will lose its charge if left standing over a period of a few months.

For periodic charging or for recharging after use, the constant-current method is generally applied, the current value needed varying somewhat with the type of cell. A common value is one-tenth of the ampere-hour capacity at the 10 h rate, i.e. 6 A for a 60 Ah battery. It is important not to overcharge. There are three methods by which the state of charge can be determined:

(i) *Colour of plates:* Fully charged cells should have clear light-grey negative plates, and rich chocolate-brown positive plates.

(ii) *Terminal voltage:* Open-circuit voltage of a fully charged cell depends on the type, being from $2 \cdot 1$ V to $2 \cdot 3$ V. If this voltage is measured with the charging current flowing, it will be increased by the voltage drop in the internal resistance of the cell.

(iii) *Specific gravity:* This is the best method of determining the state of charge, the specific gravity of the electrolyte varying from $1 \cdot 25$ for a fully charged cell to about $1 \cdot 17$ for a discharged cell. These figures apply to flooded-type storage batteries, and may vary slightly for other types of cell.

Specific-gravity readings can be taken using a **hydrometer**, shown in Figure 8.7. This consists of a wide glass tube with a rubber tube fitted to one end, and a rubber bulb to the other. The rubber tube is inserted into the electrolyte after squeezing the bulb, release of which will draw a small quantity of acid into the glass tube. This tube contains a small glass float. In a dense acid, the float is lifted high, but sinks lower in a weaker electrolyte: direct readings can be taken from graduations on the float neck.

Figure 8.7   Glass-bulb hydrometer

Acid temperature will affect specific-gravity readings. The values given are for 15°C. For each degree Celsius above or below this figure, $0 \cdot 0007$ should be added or subtracted, respectively.

'Trickle' and 'float' charging

Nickel–cadmium cells are usually float charged during use at a constant voltage, set to minimise the resulting overcharge current. This is done to keep them in optimum condition and to avoid the need to have a larger than necessary battery to take into account self-discharge.

A lead–acid cell can be kept in a healthy condition for long periods by making good the losses due to self discharge as they occur. This is done by continually trickle charging with a small current, the value of which in milliamperes is equal to the ampere-hour capacity for cells of up to 100 Ah (i.e. 60 mA for 60 Ah battery). Above this capacity, the correct trickle-charging rate is given by

$$\{70 + (3 \times 10 \text{ hour capacity})\}\text{milliamperes}$$

Some DC systems use 'floating' batteries, of the same nominal voltage as the supply and connected directly across it. If the battery voltage falls, supply voltage is greater, and the battery charges until values are equal. If the supply fails, the battery at once takes over.

Precautions

Precautions must be carefully taken where lead–acid cells are used or charged, mainly because the cells 'gas', giving off hydrogen and oxygen during charge and discharge. The following are the most important points to remember:

(i) Keep the electrolyte in flooded cells at the correct level with distilled water, to make good the loss due to evaporation and gassing.

(ii) Use no materials or finishes which will be attacked by acid in the battery room. Spilled acid, and acid vapour given off during gassing, will quickly corrode most exposed metals other than lead. Use an asphalt floor, and coat wooden surfaces liberally with anti-acid paint.

(iii) Ventilate the battery room well, if necessary using corrosion-proofed fans.

(iv) Do not allow a naked flame in the room; and prevent sparking by switching off a circuit before connecting and disconnecting. The gases present are explosive when in the correct proportions.

(v) Mop up spilled acid immediately and wash with a soda solution. Acid on clothing will quickly cause holes to appear.

(vi) Do not allow acid to enter the eyes. If it does so, immediately lie down and run clean water over the eye for as long as possible. Consult a doctor. Acid on the hands is not in itself dangerous, but can be easily transferred to more vulnerable parts of the body.

(vii) When mixing acid for the initial charge of new cells, always add acid to water, and *not* the reverse.

(viii) Watch cell temperature, as excessive heat will damage lead–acid cells. Acid temperature should not exceed 36°C.

(ix) Keep battery terminals clean and coated with petroleum jelly.

(x) Remember that a short-circuit across battery or cell terminals can result in very high currents which may cause fire or burns. Use only insulated tools and take great care not to allow terminals to be connected together inadvertently.

Alkaline cells: maintenance

These cells are lighter than the lead–acid type, have greater mechanical strength, can withstand heavy currents without damage, give off no corrosive fumes, and are not affected by being left in the discharged condition. Charge is held much better than with lead–acid cells, the makers claiming 70% capacity after three years without attention. Trickle charging is thus seldom necessary, although it can be employed when required. Charging after use, or at six-monthly intervals, is all the attention usually required.

The specific gravity of the cells does not change with condition from the value 1·19, the only indication of charge state being terminal voltage, which will normally be about 1·3 V on open circuit or 1·75 V on charge. If in doubt, continue to charge, as these cells are not damaged by overcharging. Constant-voltage or constant-current charging can be used, constant current being about twice the value for a lead–acid cell of the same capacity. Topping up with distilled water is necessary, but precautions (ii) to (vii) for lead–acid cells do not apply. Internal resistance of these cells is generally higher than the equivalent values for lead–acid cells, so voltage varies over a wider range with load.

Comparative charge/discharge characteristics for lead–acid and alkaline cells are shown in Figure 8.8.

Figure 8.8   Operational characteristics of cells
‾‾‾‾ typical lead–acid cell
---- typical alkaline cell

8.5   Internal resistance

The path taken by current as it passes through a cell will have resistance. This is the internal resistance of the cell, and is important because of the voltage drop which it causes in the cell, and which results in the terminal voltage being less than the EMF on discharge. The internal voltage drop, given by multiplying internal resistance by current, must be subtracted from the EMF to find the output voltage; that is

$$E - IR_c = U$$

where E = cell EMF, V

         I = current taken from the cell, A

         R_c = cell internal resistance, Ω

         U = cell terminal PD, V

The internal resistance of a cell usually increases with its age, and with ill treatment such as excessive discharging current, standing in the discharged condition etc.

Example 8.1

A cell has an internal resistance of $0 \cdot 02$ Ω and an EMF of $2 \cdot 2$ V. What is its terminal PD if it delivers (*a*) 1 A (*b*) 10 A or (*c*) 50 A?

(*a*) $U = E - IR_c = 2 \cdot 2 - (1 \times 0 \cdot 02)$ volts

                      $= 2 \cdot 2 - 0 \cdot 02$ volts

                      $= 2 \cdot 18$ V

(*b*) $U = E - IR_c = 2 \cdot 2 - (10 \times 0 \cdot 02)$ volts

                      $= 2 \cdot 2 - 0 \cdot 2$ volts

                      $= 2$ V

(*c*) $U = E - IR_c = 2 \cdot 2 - (50 \times 0 \cdot 02)$ volts

                      $= 2 \cdot 2 - 1 \cdot 0$ volts

                      $= 1 \cdot 2$ V

This example illustrates how the terminal PD falls off as the current increases. The effect is noticeable when a car is started at night. The heavy current taken by the starter motor reduces the battery voltage and the lights dim for as long as the starter is used.

    The internal resistance of a cell depends on its design, construction, age and condition. Internal resistance can be measured using a high-resistance voltmeter and an ammeter connected with a switch and resistor (Figure 8.9). When the switch is open, no current is taken from the cell (if we neglect the voltmeter current), so the voltmeter reads cell EMF. If the switch is closed, the terminal voltage and current are measured.

Since $U = E - IR_c$

$$R_c = \frac{E - U}{I}$$

Figure 8.9   Circuit for calculation of cell internal resistance

The resistor R is used to adjust the cell current to a convenient value which will simplify the calculation.

Example 8.2

The EMF of a cell is measured as $2 \cdot 1$ V, and its terminal PD as $1 \cdot 9$ V when it carries a current of 5 A. What is its internal resistance?

$$R_c = \frac{E - U}{I}$$

$$= \frac{2 \cdot 1 - 1 \cdot 9}{5} \text{ ohm}$$

$$= \frac{0 \cdot 2}{5} \text{ ohm}$$

$$= 0 \cdot 04 \ \Omega$$

The symbol for a cell is shown in Figure 8.10, the positive connection being represented by a long line and the negative by a shorter, thicker line. It is often convenient to show cell internal resistance as a series-connected external resistor.

When a cell or battery is on charge, the applied terminal voltage must be greater than the EMF, so that current is forced against the opposition of the EMF. Since the effective voltage is the difference between the terminal voltage and the EMF,

$$I = \frac{U - E}{R_c}$$

from which

$$R_c = \frac{U - E}{I}$$

and

$$U = E + IR_c$$

Compare this with $U = E - IR_c$ for a discharging cell.

Figure 8.10 Representation of cell in circuit diagram

Example 8.3

A cell with an EMF of 2 V and an internal resistance of $0 \cdot 08$ Ω is to be charged at 5 A. What terminal voltage must be applied?

$$U = E + IR_c$$

$$= 2 + (5 \times 0 \cdot 08) \text{ volts}$$

$$= 2 + 0 \cdot 4 \text{ volts}$$

$$= 2 \cdot 4 \text{ V}$$

Example 8.4

A cell is charged at 10 A when a terminal voltage of $2 \cdot 7$ V is applied. If the cell EMF is $2 \cdot 2$ V, what is the internal resistance?

$$R_c = \frac{U - E}{I}$$

$$= \frac{2 \cdot 7 - 2 \cdot 2}{10} \text{ ohm}$$

$$= \frac{0 \cdot 5}{10} \text{ ohm}$$

$$= 0 \cdot 05 \text{ Ω}$$

8.6 Batteries

A single cell is often incapable of providing a high enough voltage for practical purposes, so several cells are connected in series to form a battery. Six cells are shown connected in this way in Figure 8.11. The total EMF of a battery of cells connected in series is given by multiplying the number of cells by the EMF of each cell, and its internal resistance by multiplying the number of cells by the internal resistance of each cell.

Figure 8.11 Cells connected in series

Example 8.5

(*a*) A cell of internal resistance $0 \cdot 05$ Ω and EMF $2 \cdot 2$ V is connected to a $0 \cdot 95$ Ω resistor. What current will flow?

(*b*) What current will flow if the same resistor is connected to a battery of six series-connected cells of this specification?

(*a*)

$$I = \frac{E}{R + R_c}$$

$$= \frac{2 \cdot 2}{0 \cdot 95 + 0 \cdot 05} \text{ amperes}$$

$$= \frac{2 \cdot 2}{1} \text{ amperes}$$

$$= 2 \cdot 2 \text{ A}$$

(*b*)

$$\text{battery EMF} = 6 \times 2 \cdot 2 \text{ volts}$$

$$= 13 \cdot 2 \text{ V}$$

$$\text{battery internal resistance} = 6 \times 0 \cdot 05 \text{ ohm}$$

$$= 0 \cdot 3 \ \Omega$$

$$I = \frac{E}{R + R_c}$$

$$= \frac{13 \cdot 2}{0 \cdot 95 + 0 \cdot 3} \text{ amperes}$$

$$= \frac{13 \cdot 2}{1 \cdot 25} \text{ amperes}$$

$$= 10 \cdot 6 \text{ A}$$

Note that the use of a battery of six cells has not increased the current six times.

Sometimes the current which can be provided by each cell for an acceptable fall in terminal voltage will be smaller than that required, and cells are then connected in parallel. It is important to notice that cells must never be parallel-connected unless they are identical in terms of EMF and internal resistance. Discrepancies in EMF will result in internal circulating currents in the battery, or unequal loadsharing, both these faults causing rapid deterioration of healthy cells. The EMF of a parallel-connected battery is that of each cell, whereas its internal resistance is the resultant resistance of the individual internal resistances connected in parallel. The arrangement is shown in Figure 8.12.

Figure 8.12 Cells connected in parallel

Example 8.6

A cell of EMF $1 \cdot 6$ V and internal resistance of $0 \cdot 3$ Ω is connected to a $0 \cdot 1$ Ω resistor. What current flows? Find the current if six of these cells are connected in parallel to the same load.

$$I = \frac{E}{R + R_c}$$

$$= \frac{1 \cdot 6}{0 \cdot 1 + 0 \cdot 3} \text{ amperes}$$

$$= \frac{1 \cdot 6}{0 \cdot 4} \text{ amperes}$$

$$= 4 \text{ A}$$

For the battery, $\qquad E = 1 \cdot 6$ V

and internal resistance $R_c = \dfrac{0 \cdot 3}{6}$ ohms

$$= 0 \cdot 05 \ \Omega$$

Therefore $\qquad I = \dfrac{E}{R + R_c}$

$$= \frac{1 \cdot 6}{0 \cdot 1 + 0 \cdot 05} \text{ amperes}$$

$$= \frac{1 \cdot 6}{0 \cdot 15} \text{ amperes}$$

$$= 10 \cdot 7 \text{ A}$$

Where high voltage *and* high current are necessary, series–parallel arrangements of cells are used, connected as shown in Figure 8.13. All

Figure 8.13   Cells connected to load in series–parallel

cells must be identical, and the EMF will be found by multiplying the EMF per cell by the number of cells connected in series. Overall internal resistance will be given by the expression

$$\frac{\text{resistance of each cell} \times \text{number of cells in series}}{\text{number of parallel groups}}$$

Example 8.7

Eighteen cells, each of EMF $2 \cdot 4$ V and internal resistance $0 \cdot 05$ Ω, are connected in three banks of six cells in series. The three banks are then connected in parallel with each other and with a resistor of $1 \cdot 9$ Ω. Find the current flow in the resistor.

The arrangement is shown in Figure 8.13.

$$\text{battery EMF} = 6 \times 2 \cdot 4 \text{ volts}$$

$$= 14 \cdot 4 \text{ V}$$

$$\text{battery internal resistance} = \frac{0 \cdot 05 \times 6}{3} \text{ ohm}$$

$$= 0 \cdot 1 \text{ Ω}$$

$$I = \frac{E}{R + R_c}$$

$$= \frac{14 \cdot 4}{1 \cdot 9 + 0 \cdot 1} \text{ amperes}$$

$$= \frac{14 \cdot 4}{2} \text{ amperes}$$

$$= 7 \cdot 2 \text{ A}$$

8.7 Capacity and efficiency

A battery of cells is a device for storing energy, the energy stored being known as the **capacity** of the battery. This is usually measured in **ampere hours** (Ah). Energy is given not, in fact, by multiplying together current and time, but by current × time × voltage, so this method of measurement is not strictly correct. If, however, a battery is capable of providing a current of 5 A for 10 h, it is said to have a capacity of 50 Ah. Such a battery could not be expected to provide, say, a current of 10 A for 5 h, since capacity decreases as the current taken increases. Capacity is therefore based on a definite discharge time, usually 10 h when the capacity is quoted **at the 10 h rate**.

Losses occur both in converting electrical energy to chemical energy (charging) and in the reverse operation of discharging. No cell can thus be 100% efficient, the actual efficiency being given, as usual, by

$$\text{efficiency} = \frac{\text{output}}{\text{input}} \times 100\%$$

There are two methods of measuring the efficiency of a cell.

$$\text{ampere-hour efficiency} = \frac{\text{output (discharge), Ah} \times 100\%}{\text{input (charge), Ah}}$$

A cell in good condition is likely to have an ampere-hour efficiency in the region of 80%, which is high since it assumes that charging and discharging terminal voltages are the same. In fact, charging terminal voltage is the sum of cell EMF and internal voltage drop, whereas discharging terminal-voltage drop is given by their difference. A theoretically truer method of calculating efficiency is the watt-hour method:

$$\text{watt-hour efficiency} = \frac{\text{average discharge, watt hours} \times 100\%}{\text{average charge, watt hours}}$$

Watt-hour efficiency for a cell in good condition is likely to be in the region of 65%.

Example 8.8

A discharged 12 V battery is charged for 10 h at 12 A, the average charging terminal voltage being 15 V. When connected to a load, a current of 10 A for 9 h at an average terminal voltage of 12 V discharges the battery. Calculate (*a*) the ampere-hour efficiency and (*b*) the watt-hour efficiency.

(*a*) $\qquad$ ampere-hour efficiency $= \dfrac{10 \times 9 \times 100\%}{10 \times 12}$

$$= 75\%$$

(*b*) $\qquad$ watt-hour efficiency $= \dfrac{10 \times 9 \times 12 \times 100\%}{10 \times 12 \times 15}$

$$= 60\%$$

8.8   Summary of formulas for Chapter 8

Discharging:

$$U = E - IR_c \qquad E = U + IR_c \qquad R_c = \frac{E - U}{I}$$

Charging:

$$U = E + IR_c \qquad E = U - IR_c \qquad R_c = \frac{U - E}{I}$$

where U = cell or battery terminal voltage, V
E = cell or battery EMF, V
R_c = cell or battery internal resistance, Ω
I = charge or discharge current, A

For n identical cells in series,

internal resistance = nR_c where R_c = internal resistance of each cell

EMF = nE where E = EMF of each cell

For m identical cells in parallel,

$$\text{internal resistance} = \frac{R_c}{m}$$

$$\text{EMF} = E$$

For a group of m parallel sets, each of n cells in series,

$$\text{internal resistance} = \frac{nR_c}{m}$$

$$\text{EMF} = nE$$

8.9   Exercises

1 Sketch one each of any form of (*a*) primary cell and (*b*) secondary cell. Label the component parts clearly. Describe how each cell operates, and the EMF in each case.

2 (*a*) Describe a nickel–cadmium (alkaline) cell. Give the characteristic charge and discharge curves, and discuss briefly the advantages and disadvantages of this form of secondary cell.
(*b*) Make a sketch of any one form of primary cell, labelling the separate parts.

3 Make a labelled sketch showing a section through a single-cell dry battery commonly used in hand torches.

4 A cell of EMF $1 \cdot 5$ V has an internal resistance of $0 \cdot 2$ Ω. Calculate its terminal PD if it delivers a current of $0 \cdot 5$ A.

5 A cell of EMF $2 \cdot 2$ V and internal resistance $0 \cdot 05$ Ω has a $1 \cdot 05$ Ω resistor connected across its terminals. Calculate the current flow and the terminal PD of the cell.

6 The EMF of a cell is measured with a high-resistance voltmeter on open circuit, and is found to be $1 \cdot 45$ V. When a current of 1 A is drawn from the cell, the terminal PD falls to $1 \cdot 25$ V. What is the internal resistance of the cell?

7 Describe a lead–acid secondary cell. Explain briefly the changes in the cell during charge and discharge. The potential difference between the terminals of a lead–acid cell on open circuit was $2 \cdot 18$ V; when the cell was discharging at the rate of 9 A the terminal PD was $2 \cdot 02$ V. Calculate the internal resistance of the cell.

8 Describe, with sketches, a lead–acid secondary cell and state briefly the chemical changes in the cell during charge and discharge. (The chemical formulas are *not* required.) Explain the importance of a low internal resistance. A lead–acid cell discharging at the rate of 6 A has a terminal PD of $1 \cdot 95$ V. On open circuit, the PD is $2 \cdot 1$ V. Calculate the internal resistance of the cell.

9 The electrolyte in a lead–acid cell is. . . . .

10 A battery consists of six 2 V cells in series. Calculate the EMF of the battery.

11 Six cells, each of EMF $1 \cdot 5$ V and internal resistance $0 \cdot 2$ Ω, are connected in series to a $1 \cdot 8$ Ω resistor. Calculate the current delivered, and the battery terminal voltage on load.

12 (*a*) What is the total EMF when a number of cells are connected in series?
(*b*) What is the purpose of connecting a number of cells in parallel?

13 A lead–acid battery for an electric truck has 15 series-connected cells, each with an EMF of $2 \cdot 3$ V and an internal resistance of $0 \cdot 01$ Ω. Calculate the terminal voltage when the battery delivers a current of 50 A.

14 A cell has an open-circuit terminal voltage of $2 \cdot 1$ V and an internal resistance of $0 \cdot 1$ Ω. Calculate the terminal PD of the cell on charge when the charging current is (*a*) 2 A, (*b*) 10 A.

15 (*a*) State the difference between a primary and a secondary cell.
(*b*) A battery comprises five primary cells, each cell having an EMF of $1 \cdot 1$ V and a rated current of 2 A. Calculate the EMF and the rated current of the battery when the cells are connected (i) in series, (ii) in parallel. Draw circuit diagrams for (i) and (ii).

16 A battery is made up of 12 identical cells connected in series. Each cell has an EMF of 2 V and an internal resistance of $0 \cdot 05$ Ω. What terminal voltage must be applied to the battery if a charging current of 20 A is required?

17 A battery with an EMF of 12 V charges at 10 A when 16 V is applied to it. What is the internal resistance of the battery.

18 A cell with an internal resistance of $0 \cdot 15$ Ω has a terminal PD of $1 \cdot 8$ V when charging at 5 A. What is the EMF of the cell?

19 A lead–acid battery comprises 50 cells in series, each of open-circuit EMF 2 V and internal resistance $0 \cdot 02$ Ω. Calculate the terminal voltage
   (a) when supplying a load of 10 A
   (b) when being charged at 10 A.

20 Three cells, each of EMF $1 \cdot 4$ V and internal resistance $0 \cdot 3$ Ω, are connected in parallel to a $0 \cdot 9$ Ω resistor. Calculate the current in the resistor and the battery terminal PD.

21 Twelve lead–acid cells, each of EMF $2 \cdot 1$ V and internal resistance $0 \cdot 015$ Ω, are connected in three series banks of four cells. The banks are connected in parallel to a load resistor. If a current of 20 A flows in this resistor, calculate
   (a) the resistor value
   (b) the terminal PD of the battery.

22 A battery of nine primary cells is connected
   (a) all cells in series
   (b) all cells in parallel
   (c) three sets in parallel, each set consisting of three cells in series.
   Each cell has an EMF of $1 \cdot 4$ V and an internal resistance of $0 \cdot 45$ Ω. The battery terminals are connected to a circuit of resistance $7 \cdot 2$ Ω. Calculate in each case (i) the current in the $7 \cdot 2$ Ω resistance, (ii) the voltage drop across the resistance.

23 A discharged lead–acid battery is charged at 5 A for 15 h at an average voltage of $7 \cdot 2$ V. On discharge, the battery gives 6 A for 10 h at an average PD of 6 V. Calculate the ampere-hour and watt-hour efficiencies.

8.10   Multiple-choice exercises

8M1   A secondary cell is one which
   (a) is less efficient than a primary cell
   (b) cannot be recharged when exhausted
   (c) can be recharged after use by means of a charging current
   (d) can be used only for standby purposes

8M2   The standard form of dry cell or battery is called the
   (a) wet Leclanché type   (b) zinc-chloride type
   (c) lead–acid type   (d) lithium type

8M3   The positive electrode of a zinc-chloride cell consists of
   (a) a carbon rod   (b) a zinc case
   (c) manganese dioxide   (d) plaster of paris

8M4   The terminal voltage of an alkaline cell is
   (a) 2 V   (b) $1 \cdot 25$ V
   (c) 9 V   (d) $1 \cdot 5$ V

8M5   The primary cell which is likely to be chosen for use in widely varying temperature conditions is the
   (a) lead–acid cell   (b) alkaline cell
   (c) PP9 battery   (d) Leclanché cell

8M6 A battery for use in an emergency-lighting system is likely to consist of
(a) 12 V car batteries connected in series
(b) lithium cells mounted in wooden boxes
(c) alkaline cells contained in plastics enclosures
(d) lead–acid cells contained in glass or plastics containers

8M7 The capacity of a cell or a battery is measured in
(a) ampere hours (b) litres
(c) volts (d) watts

8M8 The specific gravity of the electrolyte in a lead–acid cell is measured using a
(a) voltmeter (b) measuring jug
(c) ammeter (d) hydrometer

8M9 The following is NOT one of the advantages of the alkaline secondary dry cell
(a) a much more uniform voltage than zinc-chloride types
(b) a very long shelf life
(c) being less expensive than the zinc-chloride type
(d) very good leakage protection

8M10 Naked flames, smoking or sparking contacts must be avoided in a lead–acid-battery charging room because
(a) the leads connecting the cells might catch fire
(b) the eyes of those working there will be affected
(c) the gases given off during charging can be explosive
(d) this is a standard fire-prevention drill

8M11 A 2 V cell with a terminal voltage of $1 \cdot 95$ V when delivering a current of $2 \cdot 5$ A has an internal resistance of
(a) $0 \cdot 02 \ \Omega$ (b) $0 \cdot 2 \ \Omega$ (c) $0 \cdot 78 \ \Omega$ (d) $0 \cdot 8 \ \Omega$

8M12 A 12 V battery consists of six 2 V cells each of internal resistance $0 \cdot 006 \ \Omega$. The terminal voltage of the battery when delivering a current of 50 A will be
(a) $11 \cdot 7$ V (b) $10 \cdot 2$ V (c) 10 V (d) $13 \cdot 8$ V

8M13 An emergency-lighting battery consists of fifty 2 V cells, each with an internal resistance of $0 \cdot 01 \ \Omega$. If the cells are connected in series, the total voltage and internal resistance of the battery will be
(a) 100 V and $0 \cdot 01 \ \Omega$ (b) 2 V and $0 \cdot 5 \ \Omega$
(c) 100 V and $0 \cdot 0002 \ \Omega$ (d) 100 V and $0 \cdot 5 \ \Omega$

8M14 Forty-eight lead–acid cells, each of terminal voltage $2 \cdot 1$ V and internal resistance $0 \cdot 05 \ \Omega$, are connected as a series–parallel battery of three series strings of 16 cells connected in parallel with each other. The terminal voltage and internal resistance of the battery will be
(a) $33 \cdot 6$ V and $0 \cdot 267 \ \Omega$ (b) $100 \cdot 8$ V and $2 \cdot 4 \ \Omega$
(c) $6 \cdot 3$ V and $0 \cdot 267 \ \Omega$ (d) $33 \cdot 6$ V and $0 \cdot 05 \ \Omega$

Chapter 9

Electromagnetic induction

9.1 Introduction

For many hundreds of years, scientists knew of the existence of the electric current and of the magnetic field, but the two were considered to have no connection. Thanks to the work of Oersted, Faraday and others, the two are now considered to be inseparable. In Chapter 6, it was shown that the flow of current in a conductor gave rise to a magnetic field. In this chapter, it will be shown that, under certain conditions, a magnetic field can be responsible for the flow of an electric current.

This effect is known as electromagnetic induction.

9.2 Dynamic induction

The word 'dynamic' suggests force and movement; in dynamic induction, a conductor is moved through a magnetic field.

If a length of flexible conductor has its ends connected to a sensitive indicating instrument, the needle of the instrument will give a sharp 'kick' when the conductor is suddenly moved across a magnetic field (Figure 9.1).

This simple experiment shows that, if a conductor moves across a magnetic field, an EMF is induced in it, and this EMF will cause a current to flow if a closed electric circuit exists. If a number of conductors are assembled so that they can rotate between the poles of a magnet, we have a simple generator, the principles of which will be further explained at a later stage.

If the flexible conductor is wound in a loop, so that the adjacent sides of two turns pass at the same speed through the same magnetic field, the deflection of the instrument will be twice as great as when the single conductor was used. From this, it follows that induced EMF depends on the length of conductor subject to the magnetic field.

If the magnet is changed for a stronger one (with greater magnetic-flux density between its polefaces), the deflection will again be greater. Thus

Figure 9.1 Induced EMF in conductor due to its movement in magnetic field

induced EMF depends on the density of the magnetic field through which the conductor passes.

The deflection of the needle can be shown to depend on the speed, or velocity, of the conductor through the magnetic field. The faster the conductor moves, the greater is the deflection.

These three effects can be shown by experiment to be related by the formula

$$e = Blv$$

where e = induced EMF, V
B = flux density of magnetic field, T
l = length of conductor in the field, m
v = velocity (speed) of the conductor, m/s

Note that e will be a steady value only as long as the conductor velocity is constant, and as long as the flux density of the magnetic field remains the same. For this reason, the symbol e, for an instantaneous value, is used. The conductor must move directly across the magnetic field, so that its path is at right angles to the magnetic flux.

Example 9.1

At what velocity must a conductor, $0 \cdot 1$ m long, cut a magnetic field of flux density $0 \cdot 8$ T if an EMF of 4 V is to be induced in it?

$$e = Blv \quad \text{so} \quad v = \frac{e}{Bl}$$

$$v = \frac{4}{0 \cdot 8 \times 0 \cdot 1} \text{ metres per second}$$

$$= 50 \text{ m/s}$$

Note that, whenever a conductor moves in a magnetic field, it will have an EMF induced in it, but that this EMF can only result in a current if there is a closed circuit. The value of such a current will depend on the EMF and the circuit resistance, following the relationship $I = E/R$, provided that E is constant.

Example 9.2

A conductor 300 mm long moves at a fixed speed of 2 m/s through a uniform magnetic field of flux density 1 T. What current will flow in the conductor

(*a*) if its ends are open-circuited?

(*b*) if its ends are connected to a 12 Ω resistor?

In both cases:

$$e = Blv$$

$$= 1 \times 0 \cdot 3 \times 2 \text{ volts}$$

$$= 0 \cdot 6 \text{ V}$$

(*a*) If the ends of the conductor are open-circuited, no current will flow.

(*b*) $$I = \frac{E}{R}$$

$$= \frac{0 \cdot 6}{12} \text{ ampere}$$

$$= 0 \cdot 05 \text{ A or } 50 \text{ mA (neglecting conductor resistance)}$$

Example 9.3

A conductor of effective length $0 \cdot 5$ m is connected in series with a milliammeter of resistance 1 Ω, which reads 15 mA when the conductor moves at a steady speed of 40 mm/s in a magnetic field. What is the average flux density of the field?

$$E = IR$$

$$= 0 \cdot 015 \times 1 \text{ volt}$$

$$= 0 \cdot 015 \text{ V}$$

$$e = Blv \text{ so } B = \frac{e}{lv}$$

Therefore $$B = \frac{0 \cdot 015}{0 \cdot 5 \times 0 \cdot 04} \text{ tesla}$$

$$= 0 \cdot 75 \text{ T}$$

An alternative method of calculating induced EMF is based on the definition of the unit of magnetic flux, the weber. If a conductor moving at constant speed cuts a total of one weber of flux in one second, the EMF induced in it will be one volt. Thus the induced EMF is equal to the average rate of cutting magnetic flux in webers per second; that is

$$\text{volts} = \text{webers/second}$$

$$\text{or } e = \frac{\Phi}{t}$$

Example 9.4

A conductor is moved across a magnetic field, having a total flux of $0 \cdot 2$ Wb, in $0 \cdot 5$ s. What will be the average EMF induced?

$$e = \frac{\Phi}{t}$$

$$= \frac{0 \cdot 2}{0 \cdot 5} \text{ volt}$$

$$= 0 \cdot 4 \text{ V}$$

9.3 Relative directions of EMF, movement and flux

If the direction of movement of the conductor in the magnetic field (Figures 9.1 and 9.2) is reversed, the EMF induced will have reversed polarity. If the magnetic-field direction is reversed, this will again reverse the polarity.

It is often necessary to be able to forecast the direction of one of the three variables (magnetic field, induced EMF and conductor movement) if the other two are known. The easiest of several methods is the application of **Fleming's right-hand (generator) rule**.

The thumb and first two fingers of the *right* hand are held mutually at

Figure 9.2 Dependence of induced EMF direction on direction of conductor movement relative to magnetic field

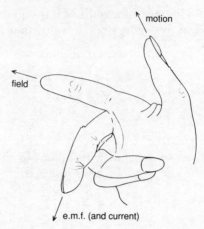

Figure 9.3   Position of right hand for application of Fleming's right-hand rule

right angles (Figure 9.3). Then, if the **F**irst-finger direction is that of the magnetic **F**ield (north pole to south pole outside the magnet), the se**C**ond-figure direction is that of **C**urrent as a result of induced EMF: then the thu**M**b shows the direction of conductor **M**ovement in the magnetic field. It is most important to ensure that the *right* hand is used.

Example 9.5

Examine the diagrams of a conductor between a pair of magnetic poles (Figure 9.4), and state

(i) the direction in which the conductor moves for Figure 9.4*a* and *b*

(ii) the polarity of the magnet system for Figure 9.4*c*

(iii) the direction of the induced EMF for Figure 9.4*d*.

Figure 9.4   Diagram for Example 9.5

The answers are

   (*a*) left to right

   (*b*) right to left

   (*c*) north pole at the top

   (*d*) upwards (out of the paper).

9.4   Simple rotating generator

We have already seen that, when a conductor moves across a magnetic field, it has an EMF induced in it. When the conductor moves *along* the field, no flux is cut, and no EMF induced. If the conductor moves at right angles to the field, flux is cut at the maximum rate and maximum EMF induced. If the conductor cuts the field obliquely, the EMF induced lies between the zero and maximum values, and depends on the angle between the line taken by the conductor and the magnetic flux.

Consider a simple rectangular loop which can be rotated between the poles of a magnet system, as shown in Figure 9.5. Connections are made to the ends of the loop by means of brushes bearing on slip rings. A section of the arrangement is shown in Figure 9.6*a*. When a conductor is in either of positions 1 and 5, it is moving along the lines of magnetic flux, but not cutting them, and so no EMF is induced in it. In positions 2, 4, 6 and 8, the conductor cuts the magnetic flux at an angle, so an EMF is induced. At positions 3 and 7, the conductor moves directly across the magnetic flux, cutting it at a maximum rate, and a maximum EMF is induced. Application of Fleming's right-hand rule will show that, as the conductor

Figure 9.5   Loop connected to slip rings and able to rotate in magnetic field

Figure 9.6   Loop rotating between poles of a magnet
     a Section through loop of Figure 9.5, which is shown in several
        positions
     b EMF in loop drawn as graph

moves from left to right through the magnetic field (positions 2, 3 and 4), the direction of the induced EMF is the opposite of that induced when the conductor passes from right to left through the magnetic field (positions 6, 7 and 8). The EMF induced in the conductor is drawn as a graph to a base of time in Figure 9.6*b*. From this, we can see that the EMF is **alternating**, that is alternately acting in opposite directions.

The opposite directions of the EMFs in opposite loop sides will drive a current through the external circuit. This current will alternate, reversing as the loop sides pass through positions 1 and 5 of Figure 9.6, and reaching maximum as the loop sides pass through positions 3 and 7.

This generator produces alternating current, and is known as an **alternator**. Alternators are used in power stations to provide electricity for the public-electricity-supply system. These very large machines have a different form of construction to our simple loop generator, but the principle is the same.

9.5   Direct-current generator

If we require a direct-current output, as we do with a DC generator, we must wire a changeover switch into circuit, and operate it to change the polarity each time the EMF reaches zero. The resulting EMF would vary continuously, but its polarity would not change. It would be physically impossible to operate such a switch by hand, because the generator revolves too quickly. What is required is an automatic switch.

Such a switch is in the form of a conducting cylinder, cut along its length and with the two halves insulated from each other. The simple **commutator** thus formed is mounted on the shaft of the machine (Figure 9.7a). A simple loop of wire forms a suitable conductor system, the two ends of the loop being connected to the two halves of the commutator. Stationary brushes rest on the commutator surface, and make contact with it, the output from this simple-loop generator being shown in Figure 9.7b.

Figure 9.8 shows the operation of the generator in simple diagrams. Figure 9.8a shows the two conductors approaching the point of maximum induced EMF. Since the individual conductor EMFs are in different directions, they add up round the loop to give a total EMF equal to twice that of each conductor. Current flows to the external load, which is connected to the loop via the commutator and brushes. In Figure 9.8b, the two conductors are approaching the point of zero EMF, and the brushes are approaching their point of changeover from one commutator segment

Figure 9.7   Simple DC generator
    a Loop connected to simple commutator and able to rotate in
      magnetic field
    b Output EMF from system

Figure 9.8   Principle of DC generator

to the next. In Figure 9.8*c*, the changeover has occurred and the lower circuit conductor is connected to rotating conductor 1, and not to conductor 2 as previously. The conductor 1 is, however, travelling through the magnetic field in a direction opposite to that which it had in Figure 9.8*a*, so its EMF has reversed to be the same as that previously induced in conductor 2. The direction of current in the external circuit is thus unchanged. The commutator then ensures that whichever conductor is passing the north pole is always connected to the negative end of the load, and the conductor passing the south pole to the positive end.

Practical DC generators have many conductor loops and commutators with a large number of segments, so the output is almost steady. The magnetic flux cut by the conductors must be as great as possible to ensure maximum induced EMF, and this is arranged by means of a **magnetic circuit**. This circuit consists of the iron polepieces, the iron body of the rotor and the frame of the machine, which is called the **yoke**. A DC machine must have any even number of poles (2, 4, 6, 8 etc.). The loops in which the EMF is induced, referred to as the **armature winding**, are let into slots in the surface of the iron cylinder forming the rotor, or rotating part of the machine. The magnetic flux is set up by field windings, which are solenoids, placed on the salient poles.

9.6   Static induction

The induced EMF (Section 9.2) was due to movement. We considered the conductor to move in a stationary field, but, of course, a similar EMF would be induced if the conductor remained still and the field moved past it. How an EMF can be induced without any physical motion at all is described here.

Consider two coils of wire placed side by side (Figure 9.9), but not touching or in electrical contact with each other. The first coil is connected in series with a battery and a switch, so that a current can be made to

Figure 9.9 Static induction

flow in it and can then be switched off. The second coil has a measuring instrument connected to its ends.

If the switch in the circuit of the first coil is operated, the instrument connected to the second coil is seen to 'kick' and then return to zero. This happens each time the switch is turned on or off, the needle moving in a different direction at each operation. A reference to Figure 9.9 shows the reason for the induced EMF. In Figure 9.9*a*, the switch is off, and the first coil sets up no magnetic flux. When the switch is on (Figure 9.9*b*), the first coil sets up a magnetic flux, some of which passes through, or 'links' with, the second coil. There has been a change in the flux linking the second coil, which has had an EMF induced in it, just as if the coil had moved into a steady magnetic field. The EMF will only be induced while the magnetic flux is changing. When the flux becomes steady, no EMF is induced.

The value of a statically induced EMF depends on the total magnetic-flux change and the time it takes to complete this change. Thus, as indicated in Figure 9.2,

$$e = \frac{\Phi}{t}$$

where e = induced EMF, V
 Φ = total magnetic-flux change, Wb
 t = time for flux change, s

Thus it is true to say that the EMF in volts induced at any instant of time is equal to the rate of change of magnetic flux at that instant in webers per second.

Example 9.6

If a current change in a coil of 200 turns induces an average EMF of 25 V, what will be the total flux change if the current takes 50 ms to complete its change?

$$e = \frac{\Phi}{t} \quad \text{so} \quad \Phi = et$$

Since the coil has 200 turns and the total induced EMF is 25 V, the induced EMF per turn will be

$$e = \frac{25}{200} = 0 \cdot 125 \text{ V}$$

$$\Phi = et$$
$$= 0 \cdot 125 \times 50 \times 10^{-3}$$
$$= 6 \cdot 25 \times 10^{-3} \text{ Wb or } 6 \cdot 25 \text{ mWb}$$

If the left-hand coil is fed from a source of alternating current, the magnetic flux set up will be continually changing and an alternating EMF will be induced in the right-hand coil. This is the principle of the transformer, which will be considered in Chapter 10.

A further study of Figure 9.9 will show that a change of magnetic-flux linkages has taken place in the left-hand coil, as well as in the right-hand coil. The left-hand coil, like the right-hand coil, will thus have an EMF induced in it, but this EMF will oppose the battery voltage and try to slow the change of current. This **self-induced EMF** is sometimes called a **back EMF**, and any circuit which has the property of inducing such an EMF in itself is said to be **self-inductive**, or just **inductive**. All circuits are to some extent, self-inductive, but some conductor arrangements give rise to a much greater self inductance than others. The unit of self inductance, which has the symbol L, is the **henry** (symbol H). The property of self inductance is discussed fully in *Electrical craft principles — Vol. 2*.

9.7   Summary of formulas for Chapter 9

$$e = Blv \qquad B = \frac{e}{lv} \qquad l = \frac{e}{Bv} \qquad v = \frac{e}{Bl}$$

where e = induced EMF, V
   B = flux density of magnetic field, T
   l = length of conductor in the field, m
   v = velocity (speed) of the conductor, m/s

$$e = \frac{\Phi}{t} \qquad \Phi = et \qquad t = \frac{\Phi}{e}$$

where e = average induced EMF, V
   Φ = total magnetic flux cut or total magnetic flux change, Wb
   t = time taken to cut flux or time taken for flux change, s

Fleming's right-hand (generator) rule

- First finger points direction of magnetic flux
- Second finger points direction of induced EMF

- Thumb points direction of conductor movement through the magnetic field.

9.8   Exercises

1 Examine the diagrams in Figure 9.10 which show a conductor being moved in a magnetic field, and state
   (*a*) the direction of induced EMF for Figure 9.10*a*
   (*b*) the direction of induced EMF for Figure 9.10*b*
   (*c*) the magnetic polarity for Figure 9.10*c*
   (*d*) the direction of conductor movement for Figure 9.10*d*.

Figure 9.10   Diagrams for Exercise 1

2 A conductor is moved at a speed of 10 m/s directly across a magnetic field of flux density 15 mT, and has an EMF of 0·3 V induced in it. What is its effective length?

3 At what speed must a conductor of effective length 180 mm be moved at right angles to a magnetic field of flux density 0·6 T to induce in it an EMF of 0·324 V?

4 A conductor of effective length 200 mm connected across a milliammeter of resistance 5 Ω is moved through a magnetic field of flux density 0·5 T. If the milliammeter reads 40 mA, at what speed must the conductor be moving?

5 What EMF will be induced in a conductor of effective length 80 mm which is moving with a velocity of 15 m/s through a magnetic field of flux density 0·4 T?

6 One conductor of a generator is 500 mm long and moves at a uniform velocity of 20 m/s in the pole flux which has an average density of 0·4 T. What is the average EMF induced in the conductor? If the winding has 200 of these conductors connected in series, what is the total generated EMF?

7 A conductor is subjected to a magnetic flux changing at the rate of 4 Wb/s. What EMF is induced in the conductor?

8 An average EMF of 1·5 V is induced in a conductor while the initial linking flux of 0·25 Wb is falling to zero. How long does the flux take to collapse?

9 A millivoltmeter connected to a conductor reads a steady 20 mV for 3 s while the conductor is subjected to a changing magnetic flux. Calculate the total flux change.

10 When a magnet is being inserted into a coil of wire, what factors govern
(*a*) the direction
(*b*) the magnitude of the induced EMF?

11 Describe with the aid of a sketch a simple loop generator which consists of a single wire loop rotating between the poles of a permanent magnet. Show the output of the loop taken from slip rings, and sketch a graph of the output voltage to a base of time.

12 Draw a diagram to show a simple two-part commutator which can be substituted for the slip rings of the generator of exercise 11. Describe how the commutator functions, and sketch a graph of the output from the machine.

13 Describe, with the aid of a sketch, the construction and action of a simple direct-current generator. State
(*a*) the factors on which the generated EMF depends
(*b*) how the generated EMF can be controlled.

9.9 Multiple-choice exercises

9M1 The word 'induction' in the electrical sense is taken to mean
(*a*) the ceremony which is held when someone joins an organisation
(*b*) the production of an EMF due to a change in linking magnetic flux
(*c*) the amount of current flowing in a resistor when a voltage is applied
(*d*) the production of magnetic flux when current flows in a coil

9M2 The value of EMF induced in a conductor when it moves through a magnetic field depends on
(*a*) the thickness of the conductor
(*b*) the total field flux and the length of conductor subject to it
(*c*) the shape of the magnetic field and the size of the conductor
(*d*) the flux density of the magnetic field, the velocity of the conductor and the length of it in the field

9M3 If a conductor wound in a loop is moved through a magnetic field
(*a*) the induced EMF will be greater than for a single conductor
(*b*) the device will be a kind of electric motor
(*c*) the induced EMF will be smaller than for a single conductor
(*d*) the individual EMFs in each loop will cancel to give no total EMF

9M4 If a conductor moving at a constant speed of 8 m/s through a magnetic field of flux density $0 \cdot 65$ T, has an EMF of $1 \cdot 3$ V induced, its length must be
(*a*) $6 \cdot 76$ m   (*b*) $0 \cdot 106$ m   (*c*) $0 \cdot 25$ m   (*d*) $0 \cdot 148$ m

9M5 A coil of 1000 turns has one side moved through a magnetic field of flux density 50 mT at a speed of $4 \cdot 2$ m/s. If 12 cm of each conductor is subjected to the flux, the total EMF induced in the coil will be
(*a*) $21 \cdot 0$ V   (*b*) $2 \cdot 7$ kV   (*c*) 21 mV   (*d*) $1 \cdot 19$ V

9M6   A coil of 450 turns carries current which sets up a total magnetic flux of 42 mWb. If an average EMF of 540 V is induced when the current is switched off, the current collapses to zero in
      (a) 50·4 ms   (b) 3·5 s   (c) 28·6 ms   (d) 35 ms

9M7   The polarity of the magnet system shown in the diagram below must be
      (a) both poles positive   (b) north pole at the top
      (c) north pole at the bottom   (d) south pole at the top

9M8   Fleming's hand rule can be used to relate the directions of the magnetic field, the current and the motion, for induced EMF. This rule can only give the correct results if we use
      (a) both hands   (b) the left hand
      (c) the right hand   (d) either hand

9M9   When using Fleming's hand rule the first finger always points the direction of the
      (a) induced EMF   (b) magnetic field
      (c) current flow   (d) force on the conductor

9M10   The EMF induced in a rectangular loop of wire which rotates in a magnetic field is
      (a) alternating   (b) very low
      (c) direct   (d) dangerous

9M11   When a conductor moves at high speed along a magnetic field so that it cuts no magnetic flux, the EMF induced in it will be
      (a) zero   (b) a maximum
      (c) alternating   (d) direct

9M12   The automatic changeover switch on the rotor of a DC machine is called the
      (a) polepiece   (b) commuter
      (c) brushes   (d) commutator

9M13   Static induction is the principle behind the operation of the
      (a) generator   (b) motor
      (c) transformer   (d) computer

Chapter 10
Basic alternating-current theory

10.1 What is alternating current?

Although direct-current systems and calculations are still indispensable to the electrical engineer, most public supplies are now alternating-current mains. The reasons for the changeover from DC to AC supplies will be considered in Section 10.2, our purpose here being to indicate how the two systems differ. The easiest method of portraying an alternating quantity is to draw a graph showing how it varies with time, as in Figure 10.1. Any part of the graph which lies above the horizontal (or zero) axis represents current or voltage in one direction, and values below it represent current or voltage in the other direction. The pattern given by the graph is known as the **waveform** of the AC system, and this usually repeats itself. There is no need for the waveform above the zero axis to have the same shape

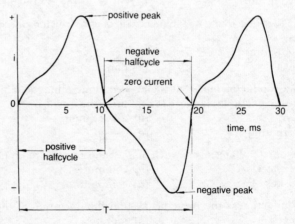

Figure 10.1   Graph of alternating current

176

as that below it although, in most AC systems derived from mains supplies, this is the case.

An alternating current is thus one which rises in one direction to a maximum value, before falling to zero and repeating in the opposite direction. Instead of drifting steadily in one direction, the electrons forming the current move backwards and forwards in the conductor.

The time taken for an alternating quantity to complete its pattern (to flow in both directions and then to return to zero) is called the **periodic time** (symbol T) for the system, which is said to complete **one cycle** in this time. The complete cycle is split into the **positive halfcycle** above the axis, and the **negative halfcycle** below it.

The number of complete cycles traced out in a given time is called the **frequency** (symbol f), usually expressed in **hertz** (Hz), which are **cycles per second** (c/s). If there are f cycles in one second, each cycle takes $1/f$ second, so that

$$T = \frac{1}{f} \quad \text{and} \quad f = \frac{1}{T}$$

Example 10.1

Calculate the frequency of the AC system shown in Figure 10.1.

From Figure 10.1 $\qquad T = 20 \text{ ms} = 0 \cdot 02 \text{ s}$

Therefore $\qquad f = \dfrac{1}{T}$

$$= \frac{1}{0 \cdot 02} \text{ hertz}$$

$$= 50 \text{ Hz}$$

A frequency of 50 Hz is the standard for the supply system in many parts of the world, including the UK, but 60 Hz systems are also common for mains supplies.

10.2 Advantages of AC systems

There are certain complications which occur when using AC supplies, which are absent with DC supplies; these complications are explained later in this chapter. However, the advantages of AC supplies have led to their general use, some of the more important being

(*a*) An alternating-current generator (often called an alternator) is more robust, less expensive, requires less maintenance, and can deliver higher voltages than its DC counterpart.

(*b*) The power loss in a transmission line depends on the square of the current carried ($P = I^2 R$). If the voltage used is increased, the current is decreased, and losses can be made very small. The simplest way of stepping up the voltage at the sending end of a line, and stepping it down again at the receiving end, is to use transformers, which will only operate efficiently from AC supplies.

(*c*) Three-phase AC induction motors are cheap, robust and easily maintained.

(*d*) Energy meters, to record the amount of electrical energy used, are much simpler for AC supplies than for DC supplies.

(*e*) Discharge lamps (fluorescent, sodium, mercury vapour etc.) operate more efficiently from AC supplies, although filament lamps are equally effective on either type of supply.

(*f*) Direct-current systems are subject to severe corrosion, which is hardly present with AC supplies.

10.3 Values for AC supplies

The alternating current or voltage changes continuously, so that it is not possible to state its value in the same simple terms that can be used for a direct current.

Instantaneous values are the values at particular instants of time, and will be different for different instants. Symbols for instantaneous values are small symbols, v for voltage, i for current and so on.

Maximum or peak values are the greatest values reached during alternation, usually occurring once in each halfcycle. Maximum values are indicated by U_m for voltage, I_m for current and so on.

Average or mean value is the average value of the current or voltage. If an average value is found over a full cycle, the positive and negative halfcycles will cancel out to give a zero result if they are identical. In such cases, it is customary to take the average value over a halfcycle. The average value of this kind of waveform can be found as shown in example 10.2. Symbols used are U_{av} for voltage, I_{av} for current and so on.

Example 10.2

Table 10.1 gives the waveform of a halfcycle of alternating voltage. Find the frequency of the supply, its instantaneous values after $1 \cdot 8$ ms and $2 \cdot 4$ ms, the maximum value and the mean value of voltage.

$$f = \frac{1}{T}$$

$$= \frac{1}{8 \text{ ms}}$$

$$= \frac{1}{\cdot 008} \text{ hertz}$$

$$= 125 \text{ Hz}$$

Table 10.1: Waveform for Example 10.2

Time, ms	0	0·25	0·5	0·75	1·0	1·25	1·5	1·75
Volts, V	0	45	72	91	104	118	142	185

Time, ms	2·0	2·25	2·5	2·75	3·0	3·25	3·5
Volts, V	240	278	295	300	280	248	195

Time, ms	3·75	4·0
Volts, V	85	0

The next step in the solution is to draw the halfcycle as a graph (Figure 10.2), reading off the instantaneous values (195 V at 1·8 ms, 287 V at 2·4 ms) and its maximum value (300 V).

To find the average or mean value, the base line (time axis) is divided into any number of equal parts. For clarity, eight parts have been chosen, although more would give greater accuracy. At the *centre* of each part, a broken line has been drawn up to the curve.

The average value of voltage will be the average length of these lines (expressed in volts). To find this, we add the voltage represented by each line and divide by the number of lines.

$$U_{av} = \frac{45 + 91 + 118 + 185 + 278 + 300 + 248 + 85}{8} \text{ volts}$$

$$= \frac{1350}{8} \text{ volts}$$

$$= 169 \text{ V}$$

Figure 10.2   Graph for Examples 10.2, 10.3 and 10.4

Effective value

Since the heat dissipated by a current is proportional to its square ($P = I^2R$), the average value of an alternating current is not the same as the direct current which produces the same heat or does the same work in the same time. The equivalent to a direct current is the value we use most in describing and calculating AC systems, and is called the **effective** or **root-mean-square** (RMS) value of the system. The RMS value is the square root of the average value of the squares of the instantaneous values. The symbols used for RMS values are U, I etc. The method of finding the effective or RMS value of a given waveform is illustrated in the following example.

Example 10.3

Find the RMS value of the voltage waveform of example 10.2.

To find the root-mean-square value

(a) divide the base into equal parts and erect a vertical line to the curve from the centre of each part (as for finding average value)
(b) square the value of each vertical line
(c) take the mean of the squared values (add them and divide by the number of lines)
(d) take the square root of the result — this is the root of the mean of the squared values.

The graph has already been drawn and vertical lines have been erected for example 10.2, and need not be repeated in this case. The sum of the squared values will be

$$45^2 + 91^2 + 118^2 + 185^2 + 278^2 + 300^2 + 248^2 + 85^2 = 294\ 468\ \text{V}^2$$

$$\text{mean of the square values} = \frac{294\ 468}{8} = 36\ 809\ \text{V}^2$$

$$\text{root-mean-square value} = \sqrt{36\ 809}\ \text{V} = 191\cdot9\ \text{V}$$

It will be seen that the RMS value is greater than the mean value, and this is always the case, except for a direct current, for which they are equal.

   Form factor for a particular waveform is the ratio of the RMS and mean values:

$$\text{form factor} = \frac{\text{RMS value}}{\text{mean value}}$$

Example 10.4

Find the form factor of the waveform of example 10.2.

$$\text{form factor} = \frac{\text{RMS value}}{\text{mean value}}$$

$$= \frac{191 \cdot 9 \text{ V}}{169 \text{ V}}$$

$$= 1 \cdot 136$$

Form factor is an indication of the shape of a waveform; the higher its value the more 'peaky' is the waveshape.

10.4   Sinusoidal waveforms

In Chapter 9, we considered a simple rectangular loop of wire rotating on an axis between the poles of a permanent magnet (Figures 9.5 and 9.6). The EMF induced in the loop is shown again in Figure 10.3, one cycle being induced for each revolution of the loop. If the loop rotates at a constant speed, the horizontal axis can be divided into equally spaced units of time as well as degrees of rotation.

The EMF induced in the loop at any instant depends on the rate of cutting lines of magnetic flux. Referring to Figure 10.4, movement from O to A induces no EMF, whereas movement from O to B induces a maximum

Figure 10.3   Sinusoidal waveform

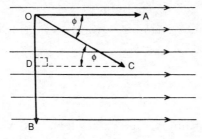

Figure 10.4   Effect of conductor direction on induced EMF

EMF, which we will call E_m. Moving the same distance from O to C at an angle of ϕ degrees to OA will induce an EMF proportional to the number of lines of magnetic flux cut, that is proportional to the length OD. OD and OC are the opposite side to ϕ and the hypotenuse, respectively, of the right-angled triangle OCD. From simple trigonometry, EMF induced in moving from O to C

$$= E_m \times \frac{OD}{OC}$$

$$= E_m \sin \phi$$

The waveshape of Figure 10.3 is thus a graph of $E_m \sin \phi$, is referred to as a 'sine wave', and is said to be 'sinusoidal' in shape. This waveshape is easily expressed as a mathematical formula, and is similar to that obtained from practical generators, so from now on we will consider all AC electrical systems to have sinusoidal waveshapes.

It is common to use the **radian** as a measure of angle; it is defined as the angle subtended at the centre of a circle by a section of the circumference of equal length to its radius. Since the circumference of a circle is $2\pi \times$ radius, there are 2π radians in $360°$, so 1 radian $= 360°/(2\pi) = 57 \cdot 3°$ approximately.

The total angular movement after t seconds of a wire loop rotating at f revolutions per second and giving an output of f cycles per second will be $2\pi ft$ radians.

Thus $\qquad\qquad\qquad e = E_m \sin 2\pi ft$

or $\qquad\qquad\qquad e = E_m \sin \omega t$

where ω (Greek letter 'omega') $= 2\pi f$ radians per second.

The average and RMS values of a sine wave are of importance. They are

$$\text{average value} = \frac{2 \times \text{maximum value}}{\pi} \text{ or } 0 \cdot 637 \times \text{maximum value}$$

$$\text{RMS value} = \frac{\text{maximum value}}{\sqrt{2}} \text{ or } 0 \cdot 707 \times \text{maximum value}$$

$$\text{form factor} = \frac{0 \cdot 707 \text{ maximum}}{0 \cdot 637 \text{ maximum}} = 1 \cdot 11$$

Values for alternating systems are always given as RMS unless otherwise stated.

Example 10.5

Find the maximum and average values for a 240 V supply.

$$\text{RMS value} = 0 \cdot 707 \times \text{maximum value}$$

Therefore
$$U_m = \frac{\text{RMS}}{0 \cdot 707}$$

$$= \frac{240}{0 \cdot 707} \text{ volts}$$

$$= 339 \text{ V}$$

$$\text{average} = 0 \cdot 637 \times U_m$$

$$= 0 \cdot 637 \times 339 \text{ volts}$$

$$= 216 \text{ V}$$

or
$$\text{average} = \frac{\text{RMS}}{\text{form factor}}$$

$$= \frac{240}{1 \cdot 11} \text{ volts}$$

$$= 216 \text{ V}$$

10.5 Phasor representation and phase difference

Wave diagrams, examples of which are shown in Figures 10.1, 10.2 and 10.3, are an extremely useful way of depicting alternating values, but they are tedious to draw exactly. An alternative method of representing an alternating quantity which varies sinusoidally is a straight line called a **phasor**, its length being proportional to the value represented. The phasor is assumed to pivot at the end without an arrowhead, and to revolve anticlockwise once for every cycle of the system it represents. Consider such a phasor rotating through a complete revolution (Figure 10.5). If the vertical height of its moving tip at various instants is transferred to a graph as shown, a sine wave results.

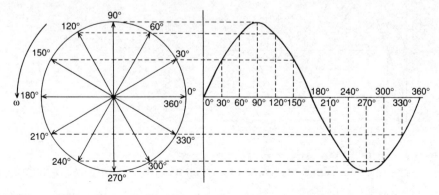

Figure 10.5   Representation of sine wave by rotating phasor

Now consider two alternating quantities (Figure 10.6a). Quantity X passes through zero, going positive 45° after quantity Y, so we say that 'X lags Y by 45°', or, alternatively, 'Y leads X by 45°'. The angle of 45° between the two quantities is the **phase angle**, and, if it is unknown, it is denoted by the symbol ϕ (small Greek letter 'phi'). The phasor diagram for the arrangement is shown in Figure 10.6b, the lengths of the phasors representing the RMS values of the alternating quantities, and the angle between them being the same as the phase angle of 45°, with Y leading X, both rotating anticlockwise.

It is often necessary to add together two alternating values. If they are **in phase** (that is, if there is no displacement of phase between them), they can be added in the same way as DC values. Thus voltages of 100 and 150 which are in phase will sum to 250 V. If the two values are not in phase, they can be added using a wave diagram or a phasor diagram, but not by simple arithmetic.

Figure 10.7(a) gives the wave diagram of a voltage U_1 of maximum value 100 V, and a voltage U_2 of maximum value 150 V which lags U_1 by 60°. The sum of these two waves at any instant is found by adding together the instantaneous values of the individual waves for that *instant*. For example

$$\text{at } 0°, \ v_1 = 0 \qquad v_2 = -130 \text{ V} \quad \text{so } v_1 + v_2 = \ -130 \text{ V}$$

$$\text{at } 30°, \ v_1 = 50 \text{ V} \qquad v_2 = \ -75 \text{ V} \quad \text{so } v_1 + v_2 = \ -25 \text{ V}$$

$$\text{at } 60°, \ v_1 = 86 \cdot 6 \text{ V} \qquad v_2 = 0 \qquad\qquad \text{so } v_1 + v_2 = +86 \cdot 6 \text{ V}$$

and so on.

These two values can be added more quickly by using a phasor diagram (Figure 10.7b). U_1 and U_2 are first drawn to scale and their **phasor sum**

Figure 10.6   Two sine waves
    a Wave diagrams of two alternating quantities, X lagging Y by 45°
    b Phasor diagram for waves of *a*

Figure 10.7 Addition of alternating values by:
  a Wave diagram
  b Phasor diagram

$U_1 + U_2$ is found by completing the parallelogram as indicated (see section 3.5 for instructions on completing the parallelogram).

Although a line whose length and direction indicate an alternating current is now called a phasor, it was for many years referred to as a **vector**, and the terms vector and vector diagram are still sometimes used.

10.6   Resistive AC circuit

If an alternating voltage of $v = U_m \sin \omega t$ is applied to a resistor, the instantaneous current

$$i = \frac{v}{R} = \frac{U_m \sin \omega t}{R}$$

and the current will be given by $i = I_m \sin \omega t$

Thus
$$I_m \sin \omega t = \frac{U_m}{R} \sin \omega t$$

so
$$I_m = \frac{U_m}{R} \text{ or, by using RMS values, } I = \frac{U}{R}$$

There is no phase difference between v and i, and the circuit calculations will be carried out in the same way as for a DC circuit. Circuit, wave and phasor diagrams are shown in Figure 10.8.

Example 10.6

A 240 V AC supply is connected to an 80 Ω resistor. Calculate the resulting current flow.

$$I = \frac{U}{R}$$

$$= \frac{240}{80} \text{ amperes}$$

$$= 3 \text{ A}$$

The current and voltage used are RMS values.

Figure 10.8   Circuit, wave and phasor diagrams for resistive AC circuit

10.7 Inductive AC circuit

The property of self inductance was considered in Chapter 9, and will be more fully discussed in Volume 2. Briefly, any coil of wire which sets up a magnetic field when it carries a current has this property, so that motor windings, coils for relays, bells and telephones possess **self inductance** (symbol L).

If current changes in such a coil, an EMF will be induced to oppose the change. If the current is increasing, the EMF will oppose the supply voltage to limit the rate of increase, and, if decreasing, will try to keep the current flowing. The unit of self inductance is the **henry** (symbol H).

Every coil of wire must possess resistance, and, because of this resistance, it is not practicable to produce a nonresistive or 'pure' inductance. However, in this section, we shall assume that a pure inductance exists, and examine the result of applying an alternating voltage to it.

Circuit, wave and phasor diagrams are shown in Figure 10.9. The EMF induced in the coil must be in opposition to the applied voltage, so, on the wave diagram, v and e are drawn with a phase displacement of 180°. The induced EMF depends on the rate of change of current $\{e = L(I_1 - I_2)/t\}$, so that, when e is zero, the rate of change of current must be zero; with a sinusoidally varying value, this only occurs at the maximum points, so current must be at maximum when e is zero. Induced EMF must also be maximum when the rate of change of current is maximum; since this maximum occurs as the current passes through zero, maximum EMF must coincide with zero current. When the current is going positive, the EMF induced must oppose this change of current, and will therefore be negative. The current wave diagram can therefore be drawn, and can be seen to **lag** the applied voltage by 90°. The phasor diagram can thus be drawn as shown, induced EMF being omitted.

We have assumed that the inductive circuit has no resistance, but, since

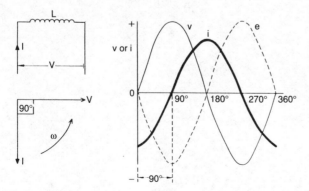

Figure 10.9 **Circuit, wave and phasor diagrams for purely inductive AC circuit**

the resulting current flow is not infinite, it must be limited by some property other than resistance. This property is called the **inductive reactance** of the coil (symbol X_L), and it can be shown that

$$X_L = \frac{U}{I} = 2\pi f L = \omega L$$

where X_L = inductive reactance of the coil, Ω
U = voltage applied to the coil, V
I = resulting current flow, A
f = supply frequency, Hz
L = coil inductance, H
$\omega = 2\pi f$

Note that, when $f = 0$, the inductive reactance will be zero. Thus the inductance of a coil has no effect on the steady flow of direct current through it, which is limited only by the coil resistance.

Example 10.7

A coil has a self inductance of $0 \cdot 318$ H and negligible resistance. Calculate its inductive reactance and the resulting current if connected to (*a*) a 240 V, 50 Hz supply; (*b*) a 100 V, 400 Hz supply.

(*a*)
$$X_L = 2\pi f L$$
$$= 2\pi \times 50 \times 0 \cdot 318 \text{ ohms}$$
$$= 100 \ \Omega$$

$$I = \frac{U}{X_L}$$

$$= \frac{240}{100} \text{ amperes}$$

$$= 2 \cdot 4 \text{ A}$$

(*b*)
$$X_L = 2\pi f L$$
$$= 2\pi \times 400 \times 0 \cdot 318 \text{ ohms}$$
$$= 800 \ \Omega$$

$$I = \frac{U}{X_L}$$

$$= \frac{100}{800} \text{ ampere}$$

$$= 0 \cdot 125 \text{ A}$$

10.8 Capacitive AC circuit

Capacitance will be treated in greater detail at the beginning of *Electrical craft principles — Vol. 2*. A capacitor is a device which is capable of storing an electric charge when a potential difference is applied to it, and can be considered to consist of two conducting plates which are very close together, but are separated by an insulator called the dielectric.

The symbol for capacitance is the letter C. The unit of capacitance is the **farad** (symbol F), but this unit is too large for most practical purposes, so the millionth part of a farad, or **microfarad** (symbol μF) is usual.

$$1\mu F = 1 \times 10^{-6} \text{ farad}$$

Smaller units of capacitance are also used, notably the nanofarad (nF) and the picofarad (pF)

$$1nF = 10^{-9} \text{ farad}$$

$$1pF = 10^{-12} \text{ farad}$$

If a direct voltage is connected to a capacitor, negatively charged electrons will be attracted from the plate connected to the positive terminal of the supply, and will be repelled from the negative terminal on to the other plate. Thus one plate will have a surplus of electrons, and the other be deficient of the same number of electrons. In this condition, the plate system is said to be **charged**, and the potential difference between the plates will increase until it equals the supply voltage, when the drift of electrons will cease.

If a capacitor is connected to a DC supply, the flow of current will die away quickly as the capacitor charges. Should, however, the supply be from an AC source, the capacitor will alternately charge and discharge with opposite polarity. Thus, although no current actually passes through the capacitor, an alternating current exists in the circuit, which can be measured by means of a suitable ammeter.

If an alternating voltage is applied to an uncharged capacitor, as the voltage passes through zero going positive, the current will immediately reach its maximum value as the capacitor starts to charge. As the charge increases, charging current will fall, reaching zero when the voltage becomes steady, which it does for an instant at its maximum value. As the voltage falls, the capacitor will discharge, and a negative current results. This pattern repeats as shown in the wave diagram of Figure 10.10, which also includes circuit and phasor diagrams. This shows us that, in a capacitive circuit, current *leads* supply voltage by 90° (compare with Figure 10.9, which shows current lagging supply voltage by 90° in a purely inductive circuit).

The flow of alternating current to a capacitor is determined by its **capacitive reactance** (symbol X_c).

$$X_c = \frac{U}{I} = \frac{1}{2\pi f C} = \frac{1}{\omega C}$$

Figure 10.10   Circuit, wave and phasor diagrams for capacitive AC circuit

where X_c = capacitive reactance, Ω
  U = supply voltage, V
  f = supply frequency, Hz
  I = circuit current, A
  C = circuit capacitance, F
  $\omega = 2\pi f$

Since the capacitance of capacitors is more often measured in microfarads, the expression can be written as

$$X_c = \frac{10^6}{2\pi f C'}$$

where C' = circuit capacitance, μF.

Example 10.8

Calculate the capacitive reactance at 50 Hz of the following capacitors: (*a*) 2 μF, (*b*) 8 μF, (*c*) 25 nF. Calculate also the current which will flow to each when connected to a 240 V, 50 Hz supply.

(*a*) $$X_c = \frac{10^6}{2\pi f C'}$$

$$= \frac{10^6}{2\pi \times 50 \times 2} \text{ ohms}$$

$$= 1590 \ \Omega$$

$$I = \frac{U}{X_c}$$

$$= \frac{240}{1590} \text{ ampere}$$

$$= 0 \cdot 151 \text{ A}$$

(b)
$$X_c = \frac{10^6}{2\pi f C'}$$

$$= \frac{10^6}{2\pi \times 50 \times 8} \text{ ohms}$$

$$= 398 \ \Omega$$

$$I = \frac{U}{X_c}$$

$$= \frac{240}{398} \text{ ampere}$$

$$= 0 \cdot 603 \ \text{A}$$

(c)
$$X_c = \frac{1}{2\pi f C'}$$

$$= \frac{10^9}{2\pi \times 50 \times 25} \text{ ohms}$$

$$= 127 \ k\Omega$$

$$I = \frac{U}{X_c}$$

$$= \frac{240}{127 \times 10^3} \text{ amperes}$$

$$= 1 \cdot 89 \ \text{mA}$$

10.9   Transformer

The transformer is undoubtedly the most important of all electrical machines. Transformers range in size from the miniature units used in some electronic applications to the huge transformers used in power stations. Although the methods of construction may differ widely, all transformers follow the same basic principles. Most transformers are required to provide an output voltage which is greater or less than that applied to the transformer.

It has been shown in Chapter 9 that, if two coils are so arranged that the magnetic flux produced by one of them links with the other, a change in the current in the first coil will result in an EMF being induced in the second coil. The EMF induced is increased if more magnetic flux is set up for a given current by the first coil, and if more of this flux can be made to link with the second coil. For this reason, the two coils are placed on a magnetic core which forms a complete magnetic circuit.

If an alternating current flows in one coil, known as the **primary**, a resulting magnetic flux will alternate in the magnetic circuit and will link with the second coil, which is known as the **secondary**. In this way, an alternating EMF will be induced in the secondary winding.

Figure 10.11 shows the simple principle of the transformer. It should be noted that there is no electrical connection between the two windings, the link being magnetic only. Because of this, the IEE Wiring Regulations (BS 7671) require one point on the secondary winding of most transformers to be earthed to prevent its potential from becoming large with respect to earth.

An EMF is induced in the secondary winding only while the current in the primary is changing. For this reason, transformers are ideally used on AC supplies. They can sometimes be used from a DC supply which is rapidly switched on and off so that the current is continually rising and falling. Figure 10.12 shows the symbol used in circuit diagrams to represent a double-wound transformer with the secondary winding and the core connected to earth.

To make the transformer as efficient as possible, all the magnetic flux produced by the primary winding should link with the secondary winding. If a transformer were made exactly as shown in Figure 10.11, a lot of the

Figure 10.11   Simplified arrangement for transformer showing main and leakage flux paths

Figure 10.12   Circuit diagram symbol for double-wound transformer

magnetic flux produced by the primary would take paths other than the path provided by the iron core.

To reduce this **leakage flux**, primary and secondary windings may be split into sections, half of each winding being placed on each side of the core. Primary and secondary may also be wound one over the other. For power transformers, the core-type construction shown in Figure 10.13 is often chosen. The windings are placed on the centre limb, the side limbs providing a path for return flux.

The EMF induced in each turn of the transformer secondary will depend on the rate of change of the magnetic flux which links with it. Since the turns of a winding are all in series with each other, the total EMF induced in a winding will be the product of the EMF per turn and the number of turns.

If leakage is neglected, the same changing magnetic flux links both primary and secondary windings, so the EMF per turn will be the same for both windings. The total induced EMF is therefore proportional to the number of turns, and the ratio of EMFs in the two windings will be the same as the ratio of their numbers of turns; that is

$$\frac{E_1}{E_2} = \frac{N_1}{N_2}$$

where E_1 = EMF induced in the primary, V
        E_2 = EMF induced in the secondary, V
        N_1 = number of primary turns
        N_2 = number of secondary turns.

In practice, both windings will have resistance, so that the voltage across the coils on load will not be quite the same as the EMFs induced in them, the differences being the voltage drops in the windings. Power-transformer windings have low resistance, and only a small error will be introduced

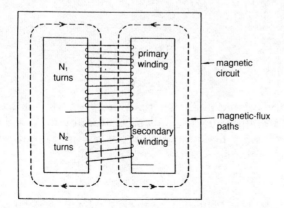

Figure 10.13   **Practical arrangement for simple transformer**

by assuming that the terminal voltage is the same as the induced EMF, so that

$$E_1 = U_1 \quad \text{and} \quad E_2 = U_2$$

where U_1 = voltage applied to the primary
U_2 = terminal voltage of the secondary

Therefore
$$\frac{U_1}{U_2} = \frac{N_1}{N_2}$$

or   voltage ratio = turns ratio

Example 10.9

A transformer is wound with 480 turns on the primary and 24 turns on the secondary. What will be the secondary voltage if the primary is fed at 240 V?

$$\frac{U_1}{U_2} = \frac{N_1}{N_2} \quad \text{so} \quad U_2 = U_1 \frac{N_2}{N_1}$$

$$U_2 = 240 \times \frac{24}{480} \text{ volts}$$

$$= 12 \text{ V}$$

Example 10.10

A transformer with a turns ratio of 2:9 is fed from a 240 V supply. What is its output voltage?

Transformer ratios are always given in the form 'primary:secondary'. This transformer has two turns on the primary for every nine on the secondary, so $N_1/N_2 = 2/9$.

$$U_2 = U_1 \frac{N_2}{N_1}$$

$$= 240 \times \frac{9}{2} \text{ volts}$$

$$= 1080 \text{ V}$$

If the secondary voltage is greater than the primary, the transformer is referred to as **stepup**, and, if smaller, as **stepdown**.

If we neglect losses, the input and output powers for a transformer will be the same. Put simply,

$$U_1 I_1 = U_2 I_2$$

where I_1 and I_2 are primary and secondary currents, respectively. Transposing, and adding the turns ratio to the expression, we have

$$\frac{U_1}{U_2} = \frac{I_2}{I_1} = \frac{N_1}{N_2}$$

Thus, if a transformer steps up the voltage from primary to secondary, the current will be less in the secondary than in the primary.

Example 10.11

A transformer has a turns ratio of 5 : 1, and is supplied at 240 V when the primary current is 2 A. Calculate the secondary current and voltage

$$U_2 = \frac{U_1 N_2}{N_1}$$

$$= \frac{240 \times 1}{5} \text{ volts}$$

$$= 48 \text{ V}$$

$$I_2 = \frac{U_1 I_1}{U_2}$$

$$= \frac{240 \times 2}{48} \text{ amperes}$$

$$= 10 \text{ A}$$

If an iron core has an alternating magnetic flux set up in it, it will get hot owing to what are called 'core losses'. These losses are subdivided into

(a) **hysteresis losses:** This subject will be considered more fully in *Electrical craft principles — Vol. 2*. Hysteresis losses can never be removed, but can be reduced by making the magnetic core of an iron containing a small percentage of silicon.

(b) **eddy-current losses:** The iron core is itself a conductor, and will have an EMF induced in it by the changing magnetic flux which it carries. The EMF causes a current to eddy back and forth in the iron, and this current causes heating.

To reduce eddy currents to a minimum, the core of a transformer is subdivided into small parts; each part has only a small share of the total induced EMF, and has high resistance owing to its small cross-section, so that eddy currents are small.

In power transformers, this subdivision is carried out by building up the core with layers of thin plates, called **laminations**. Each lamination is insulated from its neighbours by being varnished or oxidised or covered with paper on one side. The laminations must be arranged so that the magnetic flux set up is along them, and not across the insulation between them, which is nonmagnetic and would reduce the strength of the field (Figure 10.14). A coil-winding machine is used to make the transformer coils, which are then slipped over the core. Thus a core of the type shown in Figure 10.13 would be made up of E and I sections, or of U and T

Figure 10.14   Part of magnetic circuit built up of laminations
(Lamination thickness has been exaggerated)

shaped sections. Joints in the core must fit as well as possible to allow the maximum magnetic flux to be set up: laminations are interleaved at the corners to improve the effective fit.

10.10   Summary of formulas for Chapter 10

For sinusoidal waveforms:

$$\text{mean (average) value} = \frac{2}{\pi} \times \text{maximum value}$$

$$= 0 \cdot 637 \times \text{maximum value}$$

$$\text{effective (RMS) value} = \frac{1}{\sqrt{2}} \times \text{maximum value}$$

$$= 0 \cdot 707 \times \text{maximum value}$$

$$\text{form factor for any waveshape} = \frac{\text{effective value}}{\text{mean value}}$$

For purely resistive AC circuit,

$$I = \frac{U}{R} \qquad R = \frac{U}{I} \qquad U = IR$$

where U = applied voltage (effective value), V
       I = resulting current (effective value), A
       R = circuit resistance, Ω.

For a purely inductive AC circuit,

$$I = \frac{U}{X_L} \qquad X_L = \frac{U}{I} \qquad U = IX_L$$

where X_L = inductive reactance, Ω.

$$X_L = 2\pi f L \qquad f = \frac{X_L}{2\pi L} \qquad L = \frac{X_L}{2\pi f}$$

where f = supply frequency, Hz
L = circuit inductance, H

For a purely capacitive AC circuit

$$I = \frac{U}{X_c} \qquad X_c = \frac{U}{I} \qquad U = IX_c$$

where X_c = capacitive reactance, Ω.

$$X_c = \frac{1}{2\pi fC} \qquad f = \frac{1}{2\pi CX_c} \qquad C = \frac{1}{2\pi fX_c}$$

where C = circuit capacitance, F

or $$X_c = \frac{10^6}{2\pi fC'}$$

where C' = circuit capacitance, μF.

For a simple transformer, neglecting leakage,

$$\frac{U_1}{U_2} = \frac{N_1}{N_2}$$

$$U_1 = \frac{U_2N_1}{N_2} \qquad U_2 = \frac{U_1N_2}{N_1} \qquad N_1 = \frac{U_1N_2}{U_2} \qquad N_2 = \frac{N_1U_2}{U_1}$$

where U_1 = primary voltage, V
U_2 = secondary voltage, V
N_1 = number of primary turns
N_2 = number of secondary turns

$$\frac{U_1}{U_2} = \frac{I_2}{I_1}$$

$$U_1 = \frac{U_2I_2}{I_1} \qquad U_2 = \frac{U_1I_1}{I_2} \qquad I_1 = \frac{U_2I_2}{U_1} \qquad I_2 = \frac{U_1I_1}{U_2}$$

where I_1 = primary current, A
I_2 = secondary current, A.

10.11   Exercises

1 An alternating voltage completes one cycle in 1 ms. What is its frequency?

2 Explain the meaning of the term 'periodic time' as applied to an AC system. What is the periodic time of a system with a frequency of 60 Hz?

3 An alternating voltage is triangular in shape, rising at a constant rate to a maximum of 300 V in 0·01 s, and then falling to zero at a constant rate in 0·005 s. The negative halfcycle is identical in shape to the positive halfcycle.

Calculate
(a) the supply frequency
(b) the average voltage
(c) the RMS voltage
(d) the form factor.

4 A voltage is represented by a sine wave, and has a maximum of 200 V. Calculate its RMS and average or mean value.

5 Explain, with the aid of a diagram, the meaning of the following terms as applied to an alternating EMF:
(a) maximum value
(b) root-mean-square value
(c) cycle.
If the frequency of the supply is 50 Hz, what is the time taken for the completion of one cycle?

6 The positive halfcycle of a sinusoidal alternating current has the following values at 1 ms intervals:

t (ms)	0	1	2	3	4	5	6	7	8	9	
I (A)	0	7	13	19	24·5	30·5	35·5	40	44·5	48	
t (ms)	10	11	12	13	14	15	16	17	18	19	20
I (A)	50·5	52	52·5	52·5	51·5	49·5	45·5	39·5	31	15	0

Plot this curve, and hence calculate
(a) the supply frequency
(b) the mean value of the current
(c) the effective value of the current
(d) the form factor
(e) the instantaneous value of current after 27 ms.

7 Using sine tables, draw out the positive halfcycle of a sinusoidal voltage of peak value 200 V, and calculate the average voltage, the RMS voltage and the form factor.

8 A sinusoidal current has an effective value of 10 A. Calculate its average and maximum values.

9 Calculate the mean and peak voltages of a 240 V sinusoidal supply.

10 Explain the following terms applied to an alternating-current wave: (a) maximum value (b) average value (c) RMS value. If the maximum value of a sine wave is 300 A, give the average and RMS values.

11 (a) Draw *freehand* two complete cycles of a sinusoidal alternating current of peak value 100 A. Label (i) one point of current zero (ii) one complete cycle (iii) one negative peak.
(b) What is the RMS value of this sine wave?
(c) Why is the *RMS* value (*effective* value) normally used to specify the current value?

12 Draw freehand two complete cycles of a sinusoidal alternating current. Label one negative peak, one positive peak, one current zero and one positive halfcycle.
   Why is an alternating current usually specified by its RMS, or effective, value? If the peak value of a sine wave of current were 200 A, what would be its RMS value?

13 Two sinusoidal currents, each of 100 A peak value, are added. One starts at zero time, and the other a quarter cycle later.
   Draw the two waves, and the wave of their sum, and measure
   (a) the peak value of the sum
   (b) the degrees after zero time at which the sum reaches its first peak.
   The following values may be used:

θ	0°	30°	60°	90°	120°	150°	180°
sin θ	0	0·5	0·866	1	0·866	0·5	0

14 Draw a scale phasor diagram to show a current of 15 A leading a current of 10 A by 30°. Find the resultant of these phasors, and its phase relative to the 10 A current.

15 Find the resultant of two 100 V AC supplies which are connected in series but 120° out of phase (a) by phasor addition, (b) by wave addition.
   Find the phase of the resultant relative to either 100 V supply.

16 Calculate the current in a 20 Ω resistor connected to a supply at
   (a) 240 V, 50 Hz; (b) 115 V, 400 Hz; (c) 1000 V, 100 Hz; (d) 200 V, 30 Hz;
   (e) 300 V, 60 Hz.

17 Calculate the inductive reactance of the given nonresistive inductor, and the current when connected to the given supply:
   (a) 1 H connected to a 240 V, 50 Hz supply
   (b) 20 mH connected to a 415 V, 60 Hz supply
   (c) 0·15 H connected to a 100 V, 400 Hz supply.

18 Calculate the capacitive reactance of the given capacitor and the current when connected to the given supply.
   (a) 10 μF connected to a 240 V, 50 Hz supply
   (b) 20 nF connected to a 12 V, 1 kHz supply
   (c) 300 μF connected to a 3 V, 30 Hz supply
   (d) 1 pF connected to a 24 V, 50 kHz supply
   (e) 40 nF connected to a 1 V, 1 Mhz supply.

19 Make a labelled sketch of a square-cross-section core for a single-phase transformer showing its detailed construction. Describe briefly its various components and the precautions to be taken during its assembly.

20 A transformer connected to a 240 V supply has 600 primary turns, and is required to have a secondary voltage of 40 V. How many turns must be wound on the secondary?

21 A transformer with a turns ratio of 500:40 has an output of 10 V. What voltage is applied to the primary winding?

22 A transformer primary winding connected across a 415 V supply has 600 turns. How many turns must be wound as the secondary if an output of 1328 V is required?

23 Make sketches indicating current flow and resulting magnetic field when current flows in a solenoid with a steel core in the shape of a closed ring. Explain how this construction can form the basis of a transformer if supplied with alternating current.

24 Draw a sketch of a transformer showing core and coils and mark in the flux path. Explain how the transformer works.

25 A transformer is supplied at 240 V. The primary winding has 4800 turns, and takes 1 A for every 10 A delivered by the secondary. Calculate (*a*) the number of turns on the secondary, (*b*) the secondary voltage.

26 Describe, with the aid of a sketch, a simple stepup transformer and explain its action.

27 A transformer has primary and secondary voltages of 720 V and 300 V, respectively. What will be the primary current when the secondary delivers 10 A?

28 A transformer has 1500 primary turns and 75 secondary turns. What current will the secondary provide if the primary carries 5 A?

29 A transformer has 200 primary turns, and is fed at 120 V. If primary and secondary currents are measured at $1 \cdot 5$ A and 120 mA, respectively, calculate the number of secondary turns and the secondary voltage.

10.12   Multiple-choice exercises

10M1   An alternating current is best defined by a graph which will indicate the
(*a*) positive halfcycles only   (*b*) waveform
(*c*) negative halfcycles only   (*d*) RMS value

10M2   The number of cycles of an alternating system traced out each second is called the
(*a*) periodic time   (*b*) halfcycle time
(*c*) waveform   (*d*) frequency

10M3   The periodic time of an alternating-current system is the
(*a*) time taken for completion of one cycle
(*b*) time before a positive halfcycle falls to zero
(*c*) time taken for completion of a halfcycle
(*d*) usually about one second

10M4   The frequency of an alternating electrical system is measured in
(*a*) cycles (*b*) volts (*c*) hertz (*d*) seconds

10M5   If the periodic time of an alternating current system is 125 µs the frequency is
(*a*) 8 kHz (*b*) $0 \cdot 000125$ s (*c*) $0 \cdot 008$ c/s (*d*) 8 Hz

10M6   A major advantage of an alternating-current supply system over a direct-current system is that
(a) electric shocks are likely to be less severe
(b) the cost will be lower
(c) the current has many values instead of only one
(d) transformers will not operate efficiently on direct-current supplies

10M7   The highest value reached by an alternating voltage in a halfcycle is called the
(a) maximum or peak value   (b) instantaneous value
(c) root-mean-square value   (d) average value

10M8   The average value of an alternating system is taken for one halfcycle only because
(a) the calculation is impossible for a full cycle
(b) otherwise positive and negative halfcycle values will cancel
(c) there is a shock danger if a full cycle is considered
(d) if a full cycle is considered the result is the effective value

10M9   The RMS (root-mean-square) value of an alternating current is
(a) the peak value of a square-wave system
(b) seldom used because it is so difficult to calculate
(c) the effective value which is almost always given
(d) the square root of the average value

10M10 The form factor of an alternating waveform is
(a) found by dividing average value by effective value
(b) an indication of the 'peakiness' of the waveshape
(c) is the ratio of maximum and RMS values
(d) is the ratio of the maximum and average values

10M11 An alternating-voltage waveform can be represented by the formula
$$e = E_m \sin 2\pi f t$$
and is called
(a) a trigonometrical waveform   (b) a phasor
(c) a sinusoidal voltage waveform (d) a radian

10M12 A phasor is
(a) part of a three-phase system
(b) a method of representing a sinusoidally varying system
(c) the angle between alternating current and voltage
(d) always measured in radians

10M13 The angle between two alternating quantities is called
(a) the lag angle, represented by π
(b) the phase angle and is fixed at 30°
(c) the angle of lead, represented by μ
(d) the phase angle, represented by ϕ

10M14 Two phasors can be added by
(a) completing the parallelogram
(b) adding their lengths with a calculator
(c) using a wave diagram
(d) drawing them end to end and measuring the result

10M15 The RMS value of a sinusoidal system can be found
    (*a*) by multiplying its maximum value by π
    (*b*) by dividing the average value by the form factor
    (*c*) by multiplying the average value by $\sqrt{2}$
    (*d*) by dividing the maximum value by $\sqrt{2}$

10M16 If a sinusoidal voltage is applied to a pure inductor
    (*a*) the average current flow will be zero
    (*b*) the current will lead the voltage by 90°
    (*c*) the current will lag the voltage by 90°
    (*d*) current and voltage will be in phase

10M17 The effect which limits current flow in a pure inductor connected to an alternating-voltage supply is called the
    (*a*) induced EMF   (*b*) inductive reactance
    (*c*) capacitive reactance   (*d*) resistance

10M18 If a supply at 240 V, 50 Hz is connected to a 420 mH inductor, the current flow will be
    (*a*) 1·82 A   (*b*) 0·57 A   (*c*) 11·4 A   (*d*) 31·6 μA

10M19 If a sinusoidal voltage is applied to a pure capacitor
    (*a*) the average current flow will be zero
    (*b*) the current will lead the voltage by 90°
    (*c*) the current will lag the voltage by 90°
    (*d*) current and voltage will be in phase

10M20 An alternating supply of 12 V at 1 kHz is connected to a capacitor, when the current is found to be 15 mA. The capacitance of the capacitor is
    (*a*) 800 Ω   (*b*) 625 nF   (*c*) 50·3 μF   (*d*) 198 nF

10M21 The two windings on a simple transformer are called the
    (*a*) magnetic circuits   (*b*) first and second windings
    (*c*) primary and secondary   (*d*) connections

10M22 The magnetic flux produced by one transformer winding which does not link with the other is called
    (*a*) linking flux   (*b*) leakage flux
    (*c*) main flux   (*d*) useful flux

10M23 For a transformer
    (*a*) turns ratio is equal to voltage ratio
    (*b*) voltages in both windings are equal
    (*c*) turns ratio is equal to current ratio
    (*d*) primary volts per turn are twice secondary volts per turn

10M24 The voltage ratio of a transformer is 240 V : 15 V. If the primary current is 250 mA, the secondary current will be
    (*a*) 15·6 mA   (*b*) 250 mA   (*c*) 4·0 A   (*d*) 400 mA

10M25 If the secondary voltage of a transformer is greater than its primary voltage, it is called a
    (*a*) dangerous transformer   (*b*) stepup transformer
    (*c*) stepdown transformer   (*d*) voltage transformer

10M26 Eddy-current losses in a transformer are reduced by making the core using
    (*a*) laminations   (*b*) solid magnetic steel
    (*c*) high-hysteresis-loss material   (*d*) two windings

Chapter 11

Electrical motor principle

11.1 Introduction

In Chapter 9, it was explained that an electric current gives rise to a magnetic field. If a second current-carrying conductor is placed in such a field, it is subjected to an electromagnetic force. This force can be used to drive the conductor, and an electric motor results.

It is interesting to reflect how much our civilisation depends on the electromagnetic principles of the generator and the motor. Without the machine, the world would be a very different place.

11.2 Force on a current-carrying conductor lying in a magnetic field

If a conductor lies at right angles to a magnetic field, it experiences a force when a current passes through it. We can verify this natural law by using a piece of wire, a permanent magnet and a battery. If the wire is placed between the magnet poles and its ends momentarily connected across the battery terminals, it will jump from its position.

Figure 11.1 explains why this force occurs. Figure 11.1*a* shows the magnetic field due to a conductor, drawn in cross-section, which is carrying a current into the plane of the paper. Figure 11.1*b* shows the magnetic field due to the magnetic poles, between which the conductor is situated, when the conductor carries no current. Since lines of magnetic flux never cross, the two fields cannot exist simultaneously in their individual forms, and the resultant field takes up the shape shown in Figure 11.1*c*. The stronger field to the right of the conductor tries to contract, and exerts a force on the conductor in much the same way as if it were a stone in a catapult. If free to do so, the conductor will move to the left. Should the conductor be moved out of the influence of the magnetic field due to the poles, it will cease to have a force applied to it.

**Figure 11.1   Force on current-carrying conductor lying in magnetic
field**

If the reader redraws the poles, conductor and magnetic field, he will
find that, if either the polarity of the magnet or the current in the conductor
is reversed, the force on the conductor will be reversed. If both are reversed,
the force remains in the same direction. Clearly, it is important to be able
to calculate the force on the conductor in given circumstances.

Experiment shows that, provided the conductor is at right angles to the
field,

$$F = BlI$$

where F = force on conductor, N
  B = flux density of magnetic field, T
  l = length of conductor in field, m
  I = current flowing in conductor, A.

Example 11.1

A conductor, $0 \cdot 2$ m long, carries a current of 25 A at right angles to a
magnetic field of flux density $1 \cdot 2$ T. Calculate the force exerted on the
conductor.

$$F = BlI$$

$$= 1 \cdot 2 \times 0 \cdot 2 \times 25 \text{ newtons}$$

$$= 6 \text{ N}$$

Example 11.2

How much current must a conductor of an electric motor carry if it is
900 mm long and is situated at right angles to a magnetic field of flux density
$0 \cdot 8$ T, if it has a force of 144 N exerted on it?

$$F = BlI$$

Therefore $I = \dfrac{F}{Bl}$

$$= \dfrac{144}{0 \cdot 8 \times 0 \cdot 9} \text{ amperes}$$

$$= 200 \text{ A}$$

11.3   Relative directions of current, force and magnetic flux

It is often important to know the direction of the force on a conductor when it carries a current of given direction in a field of given polarity. One method is to draw out the field as shown above, but there is a rule which links the directions of the current, field and force and enables us to find the third if the directions of the other two are known.

This is **Fleming's left-hand (motor) rule**. The thumb, the first finger and the second finger of the left hand are extended, so that all three are at right angles to each other (Figure 11.2). If the first finger points in the direction taken by the magnetic field, and the second finger in the direction of current flow, the thumb gives the direction of motion of the conductor as a result of the force applied to it. This is easily remembered by noticing that the **F**irst finger gives the magnetic **F**ield direction, the se**C**ond finger gives the **C**urrent direction and the thu**M**b gives the direction of conductor **M**otion as a result of the force.

A little practice will show how easy this rule is to apply, but it must be carried out using the left hand, and applies only to the motor effect.

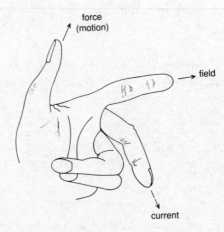

Figure 11.2   Position of left hand for application of Fleming's left-hand rule

Example 11.3

Refer to Figure 11.3, and give
  (*a*) the direction of the force on the conductor in Figure 11.3*a*
  (*b*) the polarity of the field system in Figure 11.3*b*
  (*c*) the direction of the current in Figure 11.3*c*
  (*d*) the direction of the force on the conductor in Figure 11.3*d*.

Applying Fleming's left-hand rule, or sketching the magnetic-field shapes, gives the results
  (*a*) right to left
  (*b*) north pole at the top
  (*c*) out of the paper
  (*d*) left to right.

Figure 11.3   Figures for Example 11.3

11.4   Lenz's law

Consider a conductor being moved by an external source of energy at right angles to a magnetic field (Figure 11.4). The application of Fleming's right-hand rule will show that the direction of induced EMF is such as to cause a current into the plane of the paper, if the conductor forms part of a closed circuit. We now have a current-carrying conductor situated in a magnetic field. As Fleming's left-hand rule will show, the direction of the force on this conductor will be opposite to the conductor movement.

  Thus, movement of a conductor in a magnetic field gives rise to current in such a direction as to oppose the movement causing it. Lenz's law states that the direction of an induced EMF is always such that it tends to cause a current which opposes the change inducing the EMF. In the case shown, extra energy must be used to overcome the reverse force; the work needed to overcome it will increase as the current increases, so that, for a generator, we have to put more mechanical energy in to get more electrical energy out.

Figure 11.4   Differing directions of conductor movement and force (Lenz's law)

The reverse force will not, of course, completely stop the conductor. If it did so, the induced EMF, and hence the current producing the reverse force, would disappear.

The law also affects induction in a circuit, which is due to a change in linking magnetic flux (Chapter 9). The EMF induced by the changing current will always be in such a direction as to resist that change. If the current is reducing, the EMF will be in the same direction as the current, and will try to maintain it; if the current is increasing, the EMF will oppose it and try to prevent the increase.

11.5   Direct-current motor principles

The direct-current motor is basically the same as the direct-current generator, which was considered in Section 9.5. Both machines are energy convertors. The generator is supplied with mechanical energy, and gives out most of this energy in electrical form. The motor takes in electrical energy and provides mechanical work.

Consider the simple rectangular-loop system shown in Figure 11.5. This is the same as the generator arrangement of Figure 9.7a but, instead of providing electricity, it must be supplied with electricity, so a DC supply is connected to the brushes.

Figure 11.6 shows the directions of the forces experienced by the conductors, which can be verified by application of Fleming's left-hand rule. The commutator reverses the current flow in a conductor as it passes from one pole to the next, so that the current in either conductor will always be the same as it passes a given pole. The direction of the force will therefore always be the same, and the loop will rotate continuously in a given direction. At the instant when the brushes are passing over the joints in the commutator, the conductors will be moving along the lines of magnetic

Figure 11.5   Loop connected to simple commutator and able to rotate in magnetic field

Figure 11.6   Principle of DC motor

flux, and will experience no force. In practice, the speed of rotation of the loop will keep it moving until this 'dead spot' is passed. Like the generator, the practical DC motor has many loops and a multisegment commutator. As a result, the force on the machine is nearly constant, and no 'dead spot' occurs.

The construction of the DC machine was described briefly in Section 9.5.

11.6   Moving-coil instrument

In Section 11.2 we saw that the force exerted on a conductor carrying current in a magnetic field depends on the magnetic-field strength, the length of conductor in the field and the conductor current ($F = BlI$). If field strength is made constant by the use of a permanent magnet, and the conductor is in the form of a coil of fixed length, the force must depend only on the current. Thus an instrument can be made to measure the current it carries, giving a deflection depending on the force exerted on its coil, and hence on its current.

This instrument is called the **permanent-magnet moving-coil instrument**, and is used widely for current and voltage measurements. It is rather like a miniature DC motor, but, instead of having a commutator to allow continuous rotation, current is fed into the coil through hair springs which also serve to limit the angle of rotation.

In many types, the shape of the permanent magnet is similar to that shown in Figure 11.7, the magnetic circuit being completed with shaped soft-iron pole shoes and cylindrical core, so that a radial and uniform magnetic field is set up in the airgap. The coil of fine insulated wire wound on an aluminium former is pivoted to swing in this field, and will always cut it at right angles (Figure 11.8). Two phosphor-bronze hairsprings serve to make electrical connections to the moving coil, as well as limiting the coil swing (controlling torque) and returning the movement to the zero position when no current flows (restoring torque). The coil moves because, when current flows in it, it becomes a series of current-carrying conductors lying in a magnetic field. The force on the two sides of the coil turn it against the torque of the control springs, equilibrium occurring when the deflecting torque due to the current is equal and opposite to the controlling torque

Figure 11.7　General arrangement of permanent-magnet moving-coil instrument

Figure 11.8 View of permanent-magnet moving-coil movement, showing coil, pointer, spindle and springs

due to the hairsprings. A light aluminium pointer is fixed to the coil, and moves over a scale to measure the current.

With most instruments, some provision must be made to ensure that the movement comes to rest quickly at its reading, without excessive oscillation. This provision is known as **damping**. With the permanent-magnet moving-coil instrument, damping is achieved automatically, since the EMF induced in the aluminium-coil former as it swings in the magnetic field causes an eddy current to flow which, obeying Lenz's law, produces a damping force which opposes the movement producing it.

The wire of which the moving coil is made is very fine so as to reduce the weight of the moving system, and cannot carry much current. In consequence, the torque on the coil is not very great, so the whole moving system is delicately mounted on jewelled bearings, and all possible precautions are taken to cut down friction. The moving system is carefully balanced, often having balance arms with adjustable weights for this purpose. Some instruments have a taut wire suspension, which replaces the spindle, bearings and springs, making the movement frictionless.

In their standard form, these instruments are limited to an angular movement of about 120°, but special movements are available to give **circular scales**, the needle being capable of swinging through an angle of about 300°. This increased scale length gives an improvement in the accuracy of reading, or alternatively, the same scale length can be accommodated on an instrument taking up less space.

Since the torque, and hence the instrument deflection, is proportional to current, the scale is linear: that is, there are equal spaces between equal divisions. Other advantages of this instrument are its accuracy, its sensitivity to small currents, and the ease with which it can be adapted for almost any value of current or voltage. The main disadvantage of the instrument is its inability to read values of alternating current at power frequencies. These currents give rise to an alternating torque; the movement, being unable to adjust to these rapid variations, remains at the zero position. Other disadvantages are that the delicate moving system is easily damaged by rough handling, and that the fine coil will not withstand prolonged overloading.

11.7   Summary of formulas for Chapter 11

For a conductor lying at right angles to a magnetic field,

$$F = BlI \qquad B = \frac{F}{lI} \qquad l = \frac{F}{BI} \qquad I = \frac{F}{Bl}$$

where F = force on conductor, N
 B = flux density of magnetic field, T
 l = length of conductor in magnetic field, m
 I = current carried by conductor, A

11.8   Exercises

1 A force of 10 N is exerted on a conductor $1 \cdot 5$ m long when carrying a current and lying at right angles to a magnetic field of flux density $1 \cdot 5$ T. What is the current?

2 If a conductor lying at right angles to a magnetic field of flux density $0 \cdot 12$ T experiences a force of 8 N when carrying a current of 5 A, what is the effective length of the conductor in the magnetic field?

3 20 mm of a conductor carrying a current of 15 mA is situated at right angles to a magnetic field, and experiences a force of $0 \cdot 33$ mN. What is the field flux density?

4 With the aid of a sketch, show how a force is produced on a current-carrying conductor lying in a magnetic field. Directions of the current, magnetic field and force should be shown.

5 What force is experienced by a busbar 2 m long which, under fault conditions, carries a current of 20 000 A in a magnetic field of flux density 100 mT?

6 One conductor on the coil of a moving-coil instrument is 10 mm long, and experiences a force of 2 μN when carrying a certain current. The airgap flux density is $0 \cdot 8$ T. What is the coil current?

7 Figure 11.9 gives examples of a current-carrying conductor lying in a magnetic field. State
   (*a*) the direction of the force on the conductor in Figure 11.9*a*
   (*b*) the direction of the force on the conductor in Figure 11.9*b*
   (*c*) the direction of the current in Figure 11.9*c*
   (*d*) the polarity of the magnet in Figure 11.9*d*.

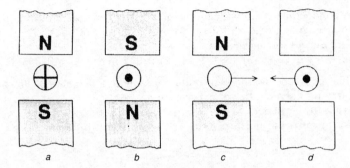

Figure 11.9   Diagrams for exercise 7

8 A moving-coil loudspeaker is required to provide a force of $0 \cdot 2$ N when its coil, which has an effective conductor length of 15 m, carries a current of 12 mA. What flux density must be set up by the magnet?

9 What length of conductor must be present in the armature of a motor which has a pole flux density of $0 \cdot 6$ T and is required to provide a turning force of 300 N when the conductor current is 25 A?

10 (*a*) Consider a single conductor carrying direct current and lying in the magnetic field between the poles of a two-pole DC motor.
      Make a sketch or diagram illustrating the above. Assuming your own directions of magnetic field and current, indicate clearly the direction in which the conductor will tend to move.
   (*b*) A conductor $0 \cdot 3$ m long lies at right angles to a magnetic field of intensity $1 \cdot 6$ T and carries a current of 25 A. Calculate the force on the conductor.

11 Sketch a two-pole DC motor showing armature, poles, commutator and brushes, and explain how a continuous torque in the same direction is obtained.

12 Sketch a moving-coil instrument and label the main parts. State two disadvantages of this type of instrument.

13 Explain why a moving-coil instrument is unsuitable for use in AC circuits.

11.9   Multiple-choice exercises

11M1   A current-carrying conductor which is subjected to a magnetic field
   (*a*) must be PVC insulated
   (*b*) will corrode more quickly than if there is no field
   (*c*) will experience a force
   (*d*) will have an EMF induced in it

11M2   The formula relating the force on a conductor (F), its length (l), the current it carries (I) and the flux density of the magnetic field to which it is subject (B) is

   (*a*) $B = Fll$ (*b*) $F = \dfrac{BI}{l}$ (*c*) $I = \dfrac{B}{Fl}$ (*d*) $F = BlI$

11M3   A conductor experiences a force of 2·5 N when 45 cm of it is subjected to a magnetic field and it is carrying a current of 65 A. The magnetic flux density of the field is
(a) 85·5 mT   (b) 73·1 T   (c) 0·855 mT   (d) 0·855 T

11M4   The figure below shows a conductor lying in a magnetic field when carrying current. It will experience a force
(a) into the paper   (b) from left to right
(c) towards the north pole (d) from right to left

11M5   The rule relating the directions of current, magnetic field and force on a conductor is
(a) Lenz's law   (b) Fleming's left-hand rule
(c) Fleming's right-hand rule   (d) Ohm's law

11M6   Lenz's law states that
(a) a conductor carrying current experiences force when subject to a magnetic field
(b) a commutator is necessary for a DC machine to work properly
(c) an induced EMF will always oppose the effect producing it
(d) when a conductor cuts a magnetic field an EMF will be induced in it

11M7   The simple electric motor shown in the figure below will
(a) rotate counter-clockwise   (b) not rotate at all
(c) rotate clockwise   (d) burn out

11M8   The major disadvantage of the permanent-magnet moving-coil instrument
is its
(*a*) inaccuracy
(*b*) nonlinear scale
(*c*) inability to measure alternating currents
(*d*) very heavy moving system

11M9   The coil of a permanent-magnet moving-coil instrument is wound with
very fine wire so that
(*a*) it can carry heavy current
(*b*) it is very light in weight
(*c*) it can easily have its range extended
(*d*) it will measure alternating and direct currents

11M10 The scale of a permanent-magnet moving-coil instrument is
(*a*) very cramped at the lower end
(*b*) difficult to read with accuracy
(*c*) reversed
(*d*) linear

Chapter 12

Practical supplies and protection

12.1 Introduction

The purpose of this chapter is to show how the basic theory already considered is put into practice to provide electrical supplies and systems.

There are a number of basic types of supply, and these will be considered in greater detail in the succeeding sections. Electricity is dangerous, as its misuse can generate heat, which results in fire. If the human body becomes part of an electric circuit, the resulting 'shock' can result in burns, or may be fatal. This chapter will begin to show the principles to be followed if these dangers are to be avoided.

12.2 Direct-current supplies

Direct-current supplies are not available from supply authorities. The reasons for this are listed in Section 10.2, which gives some of the more important advantages of AC systems.

Although mains supplies are all of the alternating-current type, it must be remembered that DC supplies are still in very wide use. For example, where an emergency supply is needed, it is often most economical to provide a battery of secondary cells. These cells give a DC supply, usually for emergency lighting in the event of a mains failure. Again, the electrical systems of motor vehicles are almost always of the DC type, so that batteries can be used for engine starting and other services (radio, parking lights etc.) which are required when the engine is not running. The modern practice is to fit an AC generator (alternator) to motor vehicles, but the output is converted to a DC system before use.

Equipment intended to break current on DC supplies is always of heavier construction (and thus more expensive) than its AC counterpart. This is because the voltage across a switch, or a similar break in a circuit, is continuous, and tends to maintain any arc that may have formed as the contacts separated. Such an arc dissipates a great deal of heat, and must be broken as soon as possible. Wide separation of switch contacts is thus required. More complicated methods of breaking the arc, such as the provision of arc chutes, or immersion of contacts in oil, are necessary in heavy-current systems.

12.3   Single-phase AC supplies

Most consumers are fed by means of a single-phase AC supply. Two wires are used, one called the **phase conductor** and usually coloured red, and the other is called the **neutral conductor** and is coloured black. Both phase and neutral conductors are called 'live' because both normally carry current. The neutral is usually earthed, the earth wire being coloured green and yellow (Section 12.5). For flexible cords, the phase conductor is coloured brown, the neutral conductor blue, and the earth conductor green and yellow.

Most houses have a single-phase AC supply, which is fed in by means of a two-core cable when the supply is underground, or two separate overhead conductors. The majority of single-phase supplies are obtained by connection to one phase of a three-phase supply. Harmonisation of supply voltages throughout Europe means that the standard UK single-phase voltage of 240 V will be reduced to 230 V.

12.4   Three-phase AC supplies

The standard method of generating, transmitting and distributing electrical energy in most countries is by use of a three-phase AC system.

If a simple wire loop is rotated in a uniform magnetic field, the EMF induced in it varies sinusoidally (Chapter 9). If three such loops are fixed together so that each makes an angle of 120° with the other two, we have an elementary system which will provide a symmetrical three-phase supply if rotated in a uniform magnetic field. A section through such a system is shown in Figure 12.1, the coils being identified by the colours red (R), yellow (Y) and blue (B). Each coil will have two ends. If the ends R_1, Y_1 and B_1 are connected together, and the ends R, Y and B are brought out to the slip rings, the coils being rotated in a clockwise direction, the outputs of the coils will be as shown in the wave diagram of Figure 12.2. The phasor diagram for the induced EMFs is also shown. It will be seen that the electrical-phase displacement between the EMFs is 120°, the same as the physical displacement of the coils. International agreements make it likely that the identification of the lines of a three-phase supply will be changed from the colours red, yellow and blue to the numbers 1, 2 and 3, respectively.

Figure 12.1   Arrangement of simple wire-loop generator to produce three-phase supply

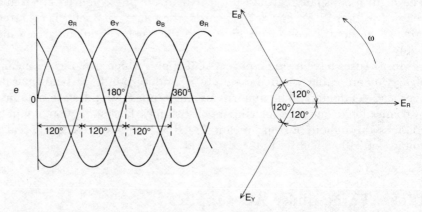

Figure 12.2   Wave and phasor diagrams for three-phase supply

There are a number of reasons for the adoption of three-phase transmission and distribution systems. Some of the most important are

(a) less copper (or aluminium) is needed for the conductors of a three-phase system which transmits a given power at a given voltage over a given distance than for a similar single-phase system
(b) three-phase motors have many advantages over single-phase motors, including smaller size, steady torque output, and the ability to self-start
(c) when connected in parallel, single-phase generators present difficulties which do not occur with three-phase generators.

Three wires, one connected to each of the slip rings of the wire loops of Figure 12.1, are necessary to carry a three-phase supply. These wires are called the **lines** and are identified by the colours red, yellow and blue or the markings L_1, L_2 and L_3. The currents flowing in the lines are called **line currents** (symbol I_L) and the PD between any two lines is called the

line voltage (symbol U_L). A fourth wire, called the **neutral**, coloured black, and usually at earth potential, is often used in conjunction with a three-phase supply. Three separate loads can be connected to a three-phase supply in either of two ways, the current in each load being known as the **phase current** (symbol I_P), and the PD across each load as the **phase voltage** (symbol U_P). The methods of connecting to a three-phase supply are known as the **delta** (or **mesh**) and **star** connections (Figure 12.3).

For the delta connection, line and phase voltages are the same, but line current is greater than phase current. For balanced systems,

$$U_L = U_P$$
$$I_L = \sqrt{3}I_P$$

delta connection

For the star connection, line and phase currents are the same, but line voltage is greater than phase voltage. For balanced systems,

$$U_L = \sqrt{3}U_P$$
$$I_L = I_P$$

star connection

These relationships will be proved in Volume 2.

Three-phase supplies are taken into large industrial and commercial premises, and are used directly to supply heavy individual loads, such as large motors, heaters, furnaces etc. If the load is unbalanced, that is if each of the three parts of the load takes a different current, a star connection with a neutral (known as a three-phase four-wire supply) is usually used. If the load is balanced, no neutral is needed, and the load may be star- or delta-connected.

For most domestic and many commercial and industrial applications, the higher voltage of the three-phase system (415 V, which is $\sqrt{3} \times 240$ V, between lines is standard; this becomes 400 V on harmonisation) is not required, and single-phase supplies are taken from the three-phase four-wire supply. Each single-phase supply is connected to one phase and the neutral, so that three separate two-wire single-phase supplies can be obtained

Figure 12.3   Balanced loads connected to three-phase supply

  a Delta or mesh connection
  b Star connection. Fourth (neutral) wire may be omitted if load
    is balanced

from a three-phase four-wire supply (Figure 12.4). These 240 (230) V supplies are used for lighting and general small-load purposes, such as cleaning, heating, ventillating etc. Some single-phase welders operate at 415 (400) V, and the diagram also shows how such a connection can be made.

Figure 12.4   Supplies obtainable from standard three-phase four-wire supply

The standard colours for the cables of a three-phase system are red, yellow and blue, with black for the neutral. For three-phase flexible cords, all three-phase conductors are brown, but marked L_1, L_2 and L_3, respectively. The neutral is then blue.

Care is taken in designing an electrical installation to 'balance the load'. To do this, the loads on the three separate phases are kept as nearly equal as possible, always bearing in mind that it may be dangerous for the areas served by different phases to overlap. This is because 415 V exists between single-phase circuits derived from different phases of a three-phase supply.

12.5   Earthing

The general mass of earth is made up almost entirely of materials which are reasonable electrical conductors themselves, or are made so by being moist. From this, it follows that a current will flow to earth through a conductor which connects a live system to earth, provided that some other point of the system at a different potential is also connected to earth.

In practice, the neutral at the supply mains is almost always connected to the general mass of earth, by connecting a conductor from the neutral at the supply origin to a metal pipe or rod driven into the ground, or to metal plates or tapes buried in it, or to the metal sheath of the supply cable, and called an **earth electrode**. Where the supply cable has a conducting sheath, it is usually connected to the neutral and to earth at the supply end, and to the consumer's earthing terminal at the incoming supply position. The complete circuit taken by the current to earth, or **earth-fault**

current, in the event of an accidental connection, or fault, to earth is shown in Figure 12.5. This is called the **earth-fault loop**.

If a human body forms part of this loop, by touching the phase conductor while also in contact with earth, the person concerned will receive a shock. The circuit shown in Figure 12.6. This possibility of shock is the disadvantage of earthing, but is comparatively easy to prevent by insulating and protecting live conductors.

One advantage of earthing is that if all the exposed metalwork of an installation is connected to earth, no significant potential difference should exist for very long between that metalwork and earth. This means that there can be no possibility of a serious shock being received from such metalwork, which includes metal conduits, cable sheaths, motor casings, metal fire reflectors and so on. The exposed metalwork is connected together and to earth by means of the **protective conductor**. This consists of the metal cable sheath, steel cable trunking or conduit, a separate earth wire, or any combination of these items. The protective conductor is connected

Figure 12.5   Earth-fault loop
The path of fault current in the loop is shown by arrows

**Figure 12.6   Electric shock owing to contact with phase conductor
and earth (direct contact)**

to earth by a conductor called the **earthing conductor**, which is attached to the incoming cable sheath or to an earth electrode. It is vitally important to ensure that the earth-continuity conductor remains intact and is always connected to earth. Figure 12.7a shows how a continuous earthing path 'shorts out' the shock path through the human body, and Figure 12.7b shows how the body may become part of the circuit if the earth path is broken.

Earthing will also help to prevent the possibility of fire due to fault. If a fault occurs between a phase conductor and earthed metalwork, there is a path for this fault current through the earth (Figure 12.7a). This extra current will cause the operation of a fuse or circuit breaker to switch off the circuit and to isolate the fault.

Figure 12.7   Shock to earth owing to earth fault
    a Protective conductor is intact, and short-circuits path through
        body if the resistance of the protective conductor is low
    b Broken protective conductor results in shock (indirect contact)

Earth-fault-loop test

The fault current mentioned above will only be large enough to operate the fuse or circuit breaker quickly if the impedance of the earth-fault loop is low enough. Fault-loop current is found by the formula

$$I = \frac{U}{Z}$$

where U is supply voltage and Z is earth-fault-loop impedance. Maximum values for the earth-fault-loop impedance for different types of protective device in different situations are given in the IEE Wiring Regulations *BS 7671* (Tables 41A1, 41A2, 41B1, 41B2, 41D, 604B1, 604B2, 605B1 and 605B2).

The impedance of the earth-fault loop cannot easily be measured using ordinary instruments, because a good deal of the loop is live while the supply is on. The supply cannot be switched off to facilitate measurement, because the supply conductors and the windings of the supply transformer are themselves part of the loop. Accordingly, a special instrument is used (see Figure 12.8). An electronic device connects a low-value resistor between phase and earth for a cycle or two of the supply voltage. The current flows for too little time to operate the protective device, but for long enough to be measured, together with the voltage, and to be used in a calculation based on $Z = U/I$. The result is displayed as the measured earth-fault-loop impedance.

Figure 12.8   Simplified circuit to a phase–earth–loop tester

Earth-electrode test

The earth electrode is the connection to the general mass of earth; it may vary from a complex grid of buried conducting plates or tapes, through metallic cable sheaths to a simple rod driven into the ground. If the earth-fault-loop impedance is too high, it is more than likely that the resistance between the electrode and the general mass of earth is too high.

A special instrument may be used for testing this resistance, or it can be tested using an AC supply, ammeter and voltmeter as shown in Figure 12.9. Two temporary test electrodes are driven, with the first electrode Y about twenty-five metres from the electrode under test, X. The second test electrode Z, is driven exactly halfway between X and Y. X is disconnected from its earthing lead, and connected as shown in Figure 12.9 so that an alternating current flows through the earth between X and Y, the value of this current being adjusted to a suitable value with R and read by the ammeter. The voltmeter reads the PD between X and Z, the latter being assumed to be outside the resistance area of both X and Y. The

Figure 12.9   Arrangements for testing an earth electrode

resistance area is the space around a live electrode in which the PD to the general mass of earth is not zero.

The earth-electrode resistance is calculated from a simple Ohm's Law relationship, i.e.

$$\text{earth-electrode resistance} = \frac{\text{voltmeter reading}}{\text{ammeter reading}}$$

To ensure that the resistance areas of the electrodes X and Y have not overlapped, which would give a false result, electrode Z is moved, first three metres nearer to $X(Z_1)$ and then three metres further from $X(Z_2)$ than its first position. If the results of all three tests are the same, the areas do not overlap. If the test results are different this indicates that the areas do overlap, and the whole test must be repeated with Y (and hence Z) further away from X.

12.6   Fuses

In the event of excessive current in a circuit, its conductors and apparatus must be protected from overheating. A fuse is the simplest device which provides this protection.

The fuse consists of a thin wire or metal tape placed in series with the circuit to be protected, so that it carries the circuit current. The wire is thick enough to carry normal load current without overheating, but, if the current exceeds the normal value, the fuse wire will melt, breaking the circuit. The thickness of the wire, or **fuse element**, is of obvious importance. If it is too thin, its resistance will be high, and the power dissipated in it by rated circuit current will raise its temperature to melting point. If it is too thick, its low resistance will dissipate little power, so that it may not melt even if the current becomes large enough to damage the circuit conductors or apparatus.

Similarly, the fuse enclosure is important. If the fuse element is open and unshrouded, it will quickly dissipate heat, and will require a large current to melt it. If enclosed so that heat cannot escape easily, it will melt at a lower current. There are two main types of fuse:

Semi-enclosed fuses

The fuse element consists of a wire of circular section which is connected to two screw terminals on the fuse carrier. The wire is usually made of an alloy containing 63% tin and 37% lead. The construction of a rewirable fuse carrier and base is shown in Figure 12.10 (rewirable is the name commonly used for this type of fuse).

Semi-enclosed fuses have the advantages of low cost and simplicity, but have many disadvantages, including

(a) aging: oxidisation of the fuse elements often leads to failure of the fuse at rated current
(b) time delay in operation: this is significant when compared with high-breaking-capacity (HBC) fuses
(c) variations in fusing current owing to the use of elements of different composition, and differing enclosures
(d) low rupturing capacity: in the event of a severe fault, the current may vaporise the element, and continue to flow in the form of an arc across the fuse terminals
(e) fire risk: the element becomes white hot as it operates
(f) the wrong fuse rating can be fitted, deliberately or by accident, by connecting an element of the wrong size: A 5 A circuit protected by a 60 A fuse in a 5 A carrier results in an obvious hazard

Figure 12.10   Semi-enclosed fuse

(g) *BS 7671* requires that the rating of a semi-enclosed fuse shall not exceed 0·725 times the current-carrying capacity of the lowest-rated conductor protected. This means that larger cables may be necessary where semi-enclosed fuses are used to provide overcurrent protection.

High-breaking-capacity fuses

This type of fuse was also known as a high-rupturing-capacity (HRC) fuse. An HBC fuse has its element enclosed in a cartridge of heatproof material, the cartridge being packed with chemically treated and graded quartz to prevent the formation of an arc. This type of fuse is nonaging, is very fast in operation, operates at a definite current for a given rating, and is capable of breaking very heavy currents. Owing to the enclosure of the element, there is no fire risk, and different current ratings are usually made with different cartridge sizes, often making it impossible to fit a fuse of the incorrect rating. The construction of a typical HBC fuse is shown in Figure 12.11.

HBC fuses are used increasingly, but their universal adoption is hampered by their high cost relative to rewirable fuses.

Figure 12.11 High-breaking-capacity fuse

Fusing factor

The current rating of a fuse is the current it will carry continuously without deteriorating. The minimum fusing current is the current which will cause the fuse to operate under given conditions in a given time. In *BS 88* the given operating time is four hours.

$$\text{fusing factor} = \frac{\text{rated minimum fusing current}}{\text{current rating}}$$

It follows that the fusing factor must exceed one; the closer it is to one, however, the less likely is it that a fault current will fail to operate the fuse. Put simply, the lower the fusing factor, the better the fuse.

The time taken for a fuse to operate is a function of the current carried.

The IEE Wiring Regulations (*BS 7671*) provide graphs of time/current characteristics for all types of fuse in Appendix 3.

Discrimination

A normal installation has a number of fuses connected in series. For example, the main fuse for a factory may be rated at 200 A, with a 60 A fuse feeding a power fuseboard which has 15 A fuses protecting individual subcircuits.

Obviously, a fault on the final subcircuit should cause the 15 A fuse to blow. If either the 60 A or the 200 A fuse blows, this will show a lack of discrimination on the part of the fuses.

Some fuses and circuit breakers are particularly fast in operation.

Care must be taken to ensure that such devices are not used to protect circuits in which much slower-operating but lower-rated fuses are used, or the main fuse or circuit breaker may operate before that in the subcircuit, putting a much wider area than necessary out of action. Such an eventuality would show bad discrimination.

12.7   Circuit breakers

The fuse element is a useful and simple protective device, but provides protection by destroying itself. After operation, it must be replaced, and demands the correct replacement HBC fuse or fusewire, the use of tools, and the expenditure of time. A particular fuse can never be tested without its self destruction, and the results of the test will not necessarily apply to the replacement fuse.

The circuit breaker is an automatic switch, which opens in the event of carrying excess current. The switch can be closed again when the current returns to normal, because the device does not damage itself during normal operation. The contacts of a circuit breaker are closed against spring pressure, and held closed by a latch arrangement. A small movement of the latch will release the contacts, which will open quickly under spring pressure to break the circuit (Figure 12.12).

The circuit breaker is so arranged that normal currents will not affect the latch, whereas excessive currents will move it to operate the breaker.

There are two basic methods by which overcurrents can operate, or 'trip', the latch:

(a) *Thermal tripping:* The load current is passed through a small heater, the temperature of which depends on the current it carries. This heater is arranged to warm a bimetal strip. This strip is made of two different metals, which are securely riveted or welded together along their length. The rate of expansion of the metals is different, so that, as the strip is warmed, it will bend and will trip the latch (Figure 12.13). The bimetal strip and heater are so arranged that normal currents

Figure 12.12   Trip mechanism of miniature circuit breaker
Broken lines show positions of components after operation of
trip lever. Even if operating dolly is held in 'on' position, trip
lever will cause separation of contacts

Figure 12.13   Principle of thermal-overload trip
a In 'set' condition
b In operated condition

will not heat the strip to tripping point. If the current increases beyond
the rated value, extra power is dissipated by the heater ($P = I^2 R$),
and the bimetal strip is raised in temperature to trip the latch.

(*b*) *Magnetic tripping:* The principle used here is the force of attraction
which can be set up by the magnetic field of a coil carrying the load
current. At normal currents, the magnetic field is not strong enough
to attract the latch, but overload currents operate the latch and trip
the main contacts (Figure 12.14).

There is always some time delay in the operation of a thermal trip, since
the heat produced by load current must be transferred to the bimetal strip.
Thermal tripping is thus best suited to small overloads of comparatively
long duration. Magnetic trips are fast acting for heavy overloads, but
uncertain in operation for light overloads. The two methods are thus
combined to take advantage of the best characteristics of each: Figure 12.15
shows a miniature circuit breaker having combined thermal and magnetic
tripping.

Figure 12.14   Types of magnetic-overload trip

a Simple attraction type. Magnetic field set up by current in flexible strip attracts strip to iron. Often used in miniature circuit breakers and combined with thermal trip (Figure 12.13)

b Solenoid type. Used on larger circuit breakers. Time lag is adjustable by varying size of oil-escape hole in dashpot piston. Current rating is adjustable by vertical movement of piston

Figure 12.15   General view of miniature circuit breaker

Circuit breakers have many advantages over fuses, but are more expensive. They are made in a wide range of sizes, from the 5 A to 100 A miniature type for domestic use, up to industrial types capable of switching thousands of amperes.

The time taken for a circuit breaker to operate is a function of the current carried. The IEE Wiring Regulations (*BS 7671*) provide graphs of time/current characteristics for all three types of miniature circuit breaker (Appendix 3).

Residual-current device

We saw in Section 12.5 that the earth-fault-loop impedance should be low enough to ensure that enough current flows in the event of a fault to blow the fuse or open the circuit breaker. In some installations, particularly where the supply is overhead and the sheath and armouring of an underground cable are not available for connection with earth, the measured impedance may be too high to ensure the operation of the fuse or circuit breaker, and a residual-current device (RCD) must be used.

A simplified arrangement is shown in Figure 12.16. The heart of the device is a magnetic core, which is wound with two main current-carrying coils, each with the same number of turns. These are connected so that the phase current in one coil provides equal and opposite ampere-turns (see Section 6.4) to the neutral current in the other coil, as long as these two currents are equal.

Any earth-fault current passes to the circuit through the phase, but does not return through the neutral. The magnetic balance in the core is lost, and an alternating flux set up in it, the value depending on the amount of imbalance, and hence on the earth-fault current.

The flux induces an EMF in the third winding, which is amplified and used to trip the breaker. A test button is required, and simulates a fault as shown.

The IEE Wiring Regulations (*BS 7671*) permit residual-current circuit breakers which operate when the earth current is as high as 2·0 A, but types operating with currents as low as 5 mA are available. RCDs rated at no more than 30 mA and operating in not more than 50 ms will provide additional protection against direct contact and hence against electric shock.

Figure 12.16   Simplified arrangement for residual-current device

The IEE Wiring Regulations (*BS 7671*) require that socket outlets must be protected by a 30 mA RCD when they are:

1. part of a TT system — that is, when no earthing terminal is provided by the electricity supply company and an earth electrode is required to earth the installation, or
2. installed in a bedroom which contains a shower cubicle, or
3. likely to feed equipment to be used out of doors, such as a lawn mower, a hedge trimmer, etc.

To prevent the possibility of the PD between earthed metal and earth becoming too high for safety, this type of protection can only be used when the operating current in amperes multiplied by the earth-loop impedance in ohms does not exceed fifty ($U = IZ$).

12.8 Risk of fire and shock

Fire

Every day there are more than 200 fires in the UK which are judged to be caused by electrical faults. The majority of these faults are due to misuse of appliances, but almost a third are due to faulty installations.

There are three main causes of these fires:

(*a*) *Overloads:* More current-using appliances are connected to a circuit than it is designed to serve, and the current exceeds the circuit rating (Figure 12.17). The circuit fuse will blow, or the circuit breaker operate, under these conditions, thus protecting the circuit. Danger arises only when the rating of the protection exceeds that of the circuit protected. For example, in Figure 12.17 three 15 A loads have been connected to a 15 A circuit, blowing the fuse. A misguided handyman may replace the fuse, and, finding that it blows again, use a 60 A fuse. The overload current will not blow this fuse, but will certainly overload the circuit conductors, rapidly aging the insulation, and causing a fire from overheating or insulation failure.

Figure 12.17 Representation of overloaded circuit

(b) *Short circuits:* If a low-resistance fault occurs between phase and neutral, a very heavy current will result. This may cause arcing at the point of the fault, with resulting fire risk. Speedy operation of the protective device will prevent the buildup of heat which will cause a fire (Figure 12.18).

Figure 12.18   Short circuit, giving very low circuit resistance, high current and causing operation of overload protection

(c) *Earth fault:* A low-resistance fault between phase and earthed metal will result in a fault current following the path indicated in Figure 12.5. Failure of the protective device to open the circuit will again result in fire risk. Since the fuse or circuit breaker is current-operated, a sufficiently large current must flow to earth to operate it, and this will depend on a low value of resistance for the current path. It is possible, if the resistance of this path (the earth-fault loop) is too high, for a current to flow which is high enough to cause overheating of the protective conductor without being high enough to blow the fuse or open the circuit breaker. Earth-fault loops are tested to ensure that their resistance is low enough to enable the supply voltage to drive sufficient current through a low-resistance fault to operate the protection.

A better method of protection will often be to fit a residual-current device (RCD) as described in Section 12.7. When the current to earth exceeds the rating of the device, the circuit will be switched off so that a possible fire risk is removed.

Shock

Section 12.12 will indicate in simple terms how an electric shock affects the human body. Over 100 deaths occur in the UK each year as a result of electric shock. In addition, large numbers of persons are burned or otherwise injured as a result of shock. These shocks occur owing to two parts of the body coming into contact with conductors at differing potentials. They can be received in three ways:

(a) *shocks from phase to neutral*: Owing to touching both conductors at the same time. Since the phase conductor is insulated and protected, such shocks are usually suffered by those foolish enough to tamper with installations, or by electricians working on live systems,

(*b*) *shocks from the phase conductor to the earth (Figure 12.6):* Owing to its insulation and protection, the phase conductor normally cannot be touched, but such accidents do happen. A not uncommon example is the dangerous practice of replacing a lamp with the supply to the lamp-holder switched on.

The IEE Wiring Regulations (*BS 7671*) refer to both these types of shock as due to 'direct contact'. Since the severity of the shock received is dependent on the voltage concerned, the shock danger can be reduced by using a lower voltage. Unfortunately, this results in increased current for a given power (*P* depends on *U* and *I*) with increased conductor-voltage drop and reduced efficiency of an appliance. One method commonly used is to employ a transformer with its secondary-winding centre tapped to earth (Figure 12.19). The voltage to earth from either conductor is then only half that of the transformer-secondary voltage.

(*c*) *shocks from noncurrent-carrying metalwork to earth:* In the event of a break in a protective conductor, the case of an appliance may be disconnected from earth. This fault is unlikely to be noticed until a second fault occurs from the phase conductor to the case, which will become live relative to earth (Figure 12.7*b*).

This type of shock is said by the IEE Wiring Regulations (*BS 7671*) to be due to 'indirect contact'.

The IEE Wiring Regulations (*BS 7671*) require that earth-fault-loop impedance shall be low enough to ensure that, in the event of a zero-impedance fault to earth, the fuse or circuit breaker must operate to disconnect the circuit within:

(i)   0·4 s for socket-outlet circuits where the equipment fed is likely to be tightly grasped, resulting in low body-contact resistance, or
(ii) 5 s for fixed equipment where body-contact resistance is likely to be higher.

Figure 12.19   Arrangement of transformers with earthed centre-tapped secondary windings

In both cases it will be impossible to prevent electric shock in the event of a fault, but its severity will be minimised by reducing the time for which it persists.

The best protection is to ensure the efficient insulation of the circuit conductors, and the continuity of the earthing system.

The use of a residual-current device with a setting of 30 mA or less will often prevent death due to shock.

12.9   Polarity

Section 12.8 has shown the danger of shock between a phase conductor and earth. If a circuit is to be switched off, or broken owing to operation of its protective device, it is ideal to open both conductors. Main switches are of the **double-pole** variety, breaking both phase and neutral conductors simultaneously.

For reasons of convenience in wiring and reduced cost, **single-pole** switches are often used in subcircuits. It is essential that such switches be connected in the phase conductor, and not the neutral of a system with

Figure 12.20   Importance of polarity in single-pole switching
 a Switch in phase conductor. Appliance safe when switch is off
 b Switch in neutral conductor. Appliance is still live although switched off

an earthed neutral. If properly connected, the part of the circuit controlled is made dead (and safe) when the switch is opened (Figure 12.20*a*). Should the supply polarity be reversed, or a single-pole switch be connected in the neutral conductor, the switch will still switch off the part of the circuit it controls. This may seem to make the circuit safe but, as Figure 12.20*b* shows, it is still live, and anyone touching a conductor (such as a fire element) will receive a shock if also in contact with earth.

For similar reasons, fuses and single-pole circuit breakers must be installed only in the phase conductors of systems with earthed neutrals. If the correct arrangement, called single-pole-and-neutral fusing, is used, the circuit is safe when the protection has operated (Figure 12.21*a*). The neutral has no fuse or circuit breaker, but contains a solid link, which may sometimes be removed for testing. Nonearthed systems sometimes use double-pole fusing, with a fuse in both phase and neutral conductors. Such a system must *not* be used where the neutral is earthed. In the event of an overload or short circuit, it is quite likely that the neutral fuse will blow, leaving the live fuse intact. The circuit will not operate in this condition, but may still be dangerous (Figure 12.21*b*).

Figure 12.21 Correct fusing for supply with earthed neutral
     a Single-pole and neutral fusing. Phase fuse 'blows' under fault conditions. No danger
     b Double-pole fusing. Neutral fuse may 'blow' under fault conditions, leaving circuit live

12.10   Safety precautions

The electrical craftsman will necessarily be called on to work on electrical circuits. In most cases, such circuits will be switched off, but the following precautions should always be taken:

(*a*) Never work on a 'live' circuit. There are situations where live working is unavoidable; in such cases work must be supervised by a fully qualified person who puts the 'permit to work' system into effect. A few minutes of inconvenience is preferable to loss of a life.

(*b*) When working on a 'live' circuit, check insulated tools carefully to ensure sound insulation.

(*c*) Never rely solely on insulated tools for protection. Wear rubber-soled shoes and see that your free hand is not in contact with earthed metal. Often, it may pay to keep your free hand in your pocket.

(*d*) Test to ensure that the circuit *really is dead* before working on it. Do not rely on the circuit markings in the fuseboard.

(*e*) If the circuit control is not close to the place of work, make sure that the supply is not restored by someone else. Keep the fuses with you, and hang a warning notice at the control position.

The Electricity at Work Regulations (1989) require that the greatest care should be taken by all who work with electricity to ensure that they are aware at all times of the need to ensure the safety of themselves and of others.

12.11   Regulations

This chapter has indicated the major hazards associated with the use of electricity, and outlined the more important methods of ensuring safety. In practice, there are many other hazards, and other regulations apply to electrical installation of various types. The major regulations are

IEE Wiring Regulations (BS 7671)

The IEE Wiring Regulations, published by the Institution of Electrical Engineers and the British Standards Institution, are the most important and far reaching of the rules applying to electrical installations. These regulations are kept up to date by issuing new editions and amendments from time to time.

The IEE Wiring Regulations (*BS 7671*) are of the greatest importance, and everyone connected with electrical installations should possess a copy and become familiar with it. Despite the length and complexity of the IEE Wiring Regulations, they cannot cover every possible contingency, and the advice and expertise of a professionally qualified electrical engineer will sometimes be required. Further guidance on good practice will be found in the relevant British Standard Codes of Practice.

Electricity Supply Regulations

Copies of these Regulations, issued by the UK Government, can be obtained from Her Majesty's Stationery Office (HMSO). They apply to the distribution and supply of electrical energy. An installation must be safe before the Regulations allow its connection to the supply; the best way of ensuring this is that it should comply fully with the IEE Wiring Regulations (*BS 7671*).

Electricity at Work Regulations (1989)

These Regulations were made under the Health & Safety at Work etc. Act (1974) and are enforced by the Health & Safety Inspectorate. The duty of everyone to consider the safety of others is stressed throughout, so there is a need to work safely at all times. The best way to ensure that users of an installation are safe is to ensure that it complies fully with the IEE Wiring Regulations (*BS 7671*).

Other special regulations

Other special regulations apply to particularly hazardous situations such as mines, quarries, and places of public entertainment. Details of full titles and sources are given in the IEE Wiring Regulations.

British Standards

Quite clearly, if an installation is to be safe for the whole of its lifetime, it must not only be installed correctly by the electrician, but also contain only materials which are safely constructed. For instance, a switch with a live metal dolly obviously would be dangerous, and no-one would make one for use at mains voltages. A switch with a live metal dolly insulated by some form of plastics would appear to be safe, but would be just as dangerous as the switch with the bare metal dolly, if the plastics insulator broke. The British Standard for such a switch would ensure that it was constructed so that the insulation would not be broken in normal service. It is more likely, however, that it would specify that any metal reinforcement used for the dolly should not become live. British Standard Specifications have been produced for nearly all types of electrical equipment, and are listed in the IEE Wiring Regulations. The Regulations themselves are now published as British Standard *7671*, giving them still greater authority, especially internationally.

Health & Safety at Work etc. Act 1974

This Act is written in general terms, and applies to many other hazards as well as electrical ones. It places a statutory duty on craftsmen to take care for their own safety as well as that of other people, and may lead to heavy fines or even imprisonment if infringed.

12.12   Electric shock

There can be few among us who have not at some time received an electric shock. The vast majority of shocks are so slight as to cause only minor discomfort, and it is this fact that has contributed to the general lack of appreciation of the real dangers involved.

An electric shock is the passage of an electric current through the body. The amount of current which is lethal varies from person to person, and also depends on the parts of the body in which it flows. To understand why we are 'shocked', we must realise that every movement we make, conscious or unconscious, is produced by muscles reacting to minute electric currents generated in the brain. These currents are distributed to the correct muscles by the 'conductors' of the nervous system.

If a current much larger than the one usually carried is forced through the nervous sytem, the muscles react much more violently than normal, and hence we experience the 'kick' associated with an electric shock. If, in addition, the nerves carry the excess current to the brain, it may destroy, or cause temporary paralysis of, the cells which generate the normal currents. Destruction of these cells means almost instant death, as the heart muscles cease to operate, and no blood is circulated. Paralysis results in unconsciousness, but, if the lung muscles are not operating, death from suffocation will follow in a few minutes.

Severity of shock

The severity of shock depends on the amount of current flowing in the body, the path it takes and the time for which it flows. As seen in Chapter 1, the amount of current flowing increases as the applied voltage increases, and decreases as circuit resistance increases. Thus, in identical circumstances, a worse shock will be received from a high voltage than from a low voltage. Many circuits, such as those for bells and telephones, are operated at low voltage so that the risk of shock is removed. High-voltage shocks are often accompanied by severe burns.

For a given voltage, the severity of the shock received will depend on circuit resistance, which is made up of the following parts:

(a) *Resistance of the installation conductors:* These form such a small proportion of the total resistance that usually they may be ignored.
(b) *Resistance of the body:* This varies considerably from person to person and, for a particular person, with time. As the body itself is made mainly of water, its resistance is quite low, but it is covered by layers of skin which have high resistance. It is in the resistance of the skin that the main variations occur. Some people have a naturally hard and horny skin which has high resistance, and others have soft, moist skin of low resistance. If the skin is wet, the moisture penetrates the pores, giving paths of low resistance. A natural result of electric shock

is for the victim to perspire profusely, thus reducing body resistance and increasing the severity of the shock.

(c) *Contact with the general mass of earth:* The body is normally separated from the conducting mass of earth by one or more layers of insulating material, i.e. shoes, floor coverings, floors etc. It is the resistance of these insulators which normally prevents a shock from being serious. For instance, a man wearing rubber-soled shoes standing on a thick carpet over a dry wood floor can touch a live conductor and feel nothing more than a slight tingle. The same man standing on a wet concrete floor in his bare feet would probably not live to describe his sensations on touching it!

This account is oversimplified, but will serve to give an indication of what occurs. Although data on the effects of shock are available (*BS PD 6519*), it would be unwise to assume that a given shock current flowing for a given time would always have certain effects or would be safe, because shock effects appear to differ from person to person and are affected by factors such as ambient temperature, health and so on. It is always best to ensure as far as possible that there will be no shock by scrupulously following the IEE Wiring Regulations (*BS 7671*). An effective backup against shock to earth is to fit a residual current device with a rating not exceeding 30 mA.

12.13   Artificial respiration

As explained in the Section 12.12, severe electric shock is often accompanied by a form of paralysis of the nervous system. The heart continues to beat, and the victim is not yet dead, but as breathing has stopped, the victim will soon die if no action is taken.

Artificial respiration keeps the patient's lungs working until the body's own system can recover and take over. It is very simple to apply, and is basically the same as that used in drowning cases. Every person should be proficient in its use.

Precautions

If you witness an accident caused by electric shock resulting in unconsciousness, or find an unconscious colleague, remember the following points:

(a) Ensure that the patient is not still in contact with the electrical system. If he or she is, and you touch the body, you may receive a shock, too. Switch off the supply, or, if this is not possible, drag the patient clear with dry clothing or some other insulator.

(b) Do not waste time trying to find out if the patient is still alive. It is vital to start respiration *at once.*

(c) Summon assistance, namely doctor, ambulance etc., at once, but do not delay artificial respiration to do so. If you are on your own, shout for help periodically.

Types of artificial respiration

For a number of years the recommended system has been that known as the **Holger–Nielsen** type. This involves lying the patient face down on the floor, alternately applying pressure to the shoulder blades and pulling forwards on the elbows to force air in and out of the lungs. This system has saved many lives, but is relatively slow, and complications can be caused if the patient has broken bones or internal injuries. Such injuries are often associated with unconsciousness owing to shock as a result of falling.

The **mouth-to-mouth** system, sometimes called the **kiss of life**, appears to be accepted generally as the quickest and most effective system.

Mouth-to-mouth artificial respiration

The patient should be quickly laid on his or her back, *with the head tilted as far backwards as possible*. This will open and straighten the air passages. Clear the mouth of foreign objects such as false teeth, vomit etc., and make sure that the tongue is not blocking the airways. Close the patient's nose by pinching the nostrils, place your mouth firmly over the patient's, and blow. When the chest is inflated, remove your mouth, drawing in breath as you do so. The patient's chest will deflate, after which the cycle should be repeated. This process should take place 10–12 times each minute, that is, once every five or six seconds. Counting the seconds to yourself may help you to prevent slowing down, which can prove fatal.

An alternative method is **mouth-to-nose** artificial respiration, which is easier in some cases. The victim's mouth should be closed with the thumb, and your mouth placed firmly over the patient's nose for the blowing operation. If the chest is not inflated, check the head position and try again. Failure on the second attempt means that you must change to the mouth-to-mouth system at once.

12.14   Summary of formulas for Chapter 12

For a delta or mesh connection,

$$U_L = U_P \qquad I_L = \sqrt{3}I_P \qquad I_P = \frac{I_L}{\sqrt{3}}$$

For a star connection,

$$I_L = I_P \qquad U_L = \sqrt{3}U_P \qquad U_P = \frac{U_L}{\sqrt{3}}$$

where U_L = line voltage (voltage between supply lines)
U_P = phase voltage (voltage across one load)
I_L = line current (current in a supply line)
I_P = phase current (current in one load)

12.15 Exercises

1 Show by means of simple diagrams what is meant by
(*a*) a single-phase supply
(*b*) a three-phase supply.

2 What is meant by the phase conductor of a two-wire AC supply?

3 Draw a sketch of the earth-fault loop, naming the parts.

4 What are the advantages and disadvantages of connecting the neutral conductor of a supply system to earth?

5 A portable drill is connected to the supply by a two-core-and-earth cable, and is controlled by a single-pole switch. The mains terminals are *phase*, *neutral* and *earth*. In the supply terminal box are a link, a fuse and an earth terminal. Make a labelled diagram of connections. Explain concisely
(*a*) why you have included the switch in a particular lead
(*b*) why a link and not a fuse is used in one lead
(*c*) how the earth-lead inclusion is intended to protect the user.

6 Describe the construction of a semi-enclosed fuse. List the advantages of this form of protection compared with an HBC fuse.

7 Show by means of a clearly labelled drawing the construction of an HBC fuse. What are the advantages of this fuse?

8 Using simple diagrams, explain the principles of operation of both thermal and magnetic overload trips. For what type of overload is each particularly suited?

9 Make a neat, labelled sketch to show the construction and operation of an adjustable solenoid-and-plunger-type overcurrent trip device.

10 What are the two major risks associated with the use of electrical energy, and how do they occur?

11 What steps are taken in the design of an electrical installation to ensure safety from fire?

12 What steps are taken in the design of an electrical installation to improve safety from shock?

13 Describe by means of circuit diagrams and brief explanations what precautions against electric shock to users are necessary for
(*a*) portable electric power tools in a factory
(*b*) lighting and heating in a domestic bathroom
(*c*) electric-bell circuits supplied from the mains.

14 With the aid of a simple sketch, explain the meaning of the term 'short circuit'.

15 Portable electrical appliances in a factory are often fed from a 240 V/110 V transformer with a centre tap on the secondary winding. Draw a diagram to show how this is connected to the appliance, and explain how it reduces the risk of electric shock.

16 Draw a diagram of a 415 V, three-phase, four-wire distribution system. Indicate on the diagram the connections for supplying the following loads:
(*a*) a single-phase lighting load
(*b*) a three-phase motor
(*c*) a 415 V, single-phase electric welder.

17 What are the standard values of the voltages available from a three-phase, four-wire, AC supply? Give the standard colour code for the four wires.

18 List the safety precautions to be taken when working on electrical circuits. What additional precautions become necessary if isolation of the circuits from the supply is not carried out?

19 Explain the action you would take if you found a colleague who was unconscious as a result of an electric shock. Assume that your colleague is still in contact with the live supply.

12.16 Multiple-choice exercises

12M1 The most important reason for the adoption of alternating-, as opposed to direct-current, systems for public electricity supplies is
(*a*) they are less likely to cause fatal electric shock
(*b*) the current alternates
(*c*) that transformers can be used, thus making transmission and distribution more efficient
(*d*) the load-carrying carbon brushes in motors are longer lasting

12M2 An electricity supply which provides phase, neutral and earth conductors is called
(*a*) a three-phase supply   (*b*) an overhead supply
(*c*) a public system   (*d*) a single-phase supply

12M3 The three-phase system is used throughout the world to transmit and distribute electrical supplies because
(*a*) the supply system needed is less expensive than other options
(*b*) three wires can be used instead of the six needed for three single-phase supplies
(*c*) the conductors used can be balanced on each side of overhead pylons
(*d*) the system is more costly than others

12M4 The phase difference between the voltages of the lines of a balanced three-phase system is
(*a*) 90°   (*b*) 180°   (*c*) 120°   (*d*) 30°

12M5 The two methods of connecting three loads to a three-phase supply are called
(*a*) delta and series connections   (*b*) star and delta connections
(*c*) right and wrong connections   (*d*) mesh and delta connections

12M6 The standard colours for the cables in a three-phase system are
(*a*) red, yellow and blue   (*b*) brown, blue and green/yellow
(*c*) red, black and green   (*d*) all black

12M7 A supply system is usually earthed
(*a*) so that there are no potential differences between conductors and earth
(*b*) so that the system will not move
(*c*) to ensure that electric shocks cannot be received
(*d*) so that there is a path for earth-fault current

12M8 An earth-fault-loop test is carried out to make sure that
(*a*) earth wires are properly connected
(*b*) enough current will flow in the event if an earth fault to open the protective device
(*c*) the earth-loop tester is working properly
(*d*) the correct testing instruments are available

12M9 When testing the resistance of an earth electrode, the distance from the auxiliary electrode to the electrode under test should usually be not less than
(*a*) 6 m   (*b*) 25 m   (*c*) 3 m   (*d*) 30 m

12M10 A fuse consists of a thin wire placed in series with the circuit to be protected so that it will
(*a*) melt and open the circuit if excessive current flows
(*b*) safely carry overload currents
(*c*) melt before the circuit design current is reached
(*d*) never break

12M11 A major advantage of the HBC (cartridge) fuse is that it
(*a*) costs more
(*b*) will get much hotter than other types when it operates
(*c*) will safely break higher currents than other types
(*d*) can be replaced by a fuse wire from a card

12M12 The term 'discrimination' in an electrical installation sense refers to
(*a*) the employment of female electricians
(*b*) the lower-rated protection device operating before the higher-rated
(*c*) selection of the correct method of earthing
(*d*) selection of the correct type of cable

12M13 A circuit breaker is often provided with two methods of tripping, which are
(*a*) thermal and magnetic   (*b*) overload and short circuit
(*c*) fast and slow   (*d*) miniature and moulded-case

12M14 A residual-current device will switch off the circuit protected when
(*a*) the total current in the circuit exceeds the rating of the device
(*b*) earth-leakage current is greater than 150 mA for 40 ms

(*c*) current to earth is greater than its residual-current rating

(*d*) circuit current exceeds 30 mA

12M15 The term 'residual current' refers to

(*a*) the total circuit current

(*b*) current left behind after the circuit is switched off

(*c*) the difference between neutral and earth currents

(*d*) the difference between the phase and the neutral currents

12M16 The two main hazards associated with the use of electrical energy are

(*a*) fire and electric shock

(*b*) power failure and high costs

(*c*) falling from a height and cuts from a hacksaw

(*d*) back damage due to lifting heavy objects and 'arc eye'

12M17 The two time periods stated in the IEE Wiring Regulations within which a circuit protective device must operate are

(*a*) 0·4 s and 5 s   (*b*) 1 s and 2 s

(*c*) 4 hours and the fusing factor   (*d*) 0·725 s and 0·4 s

12M18 Before working on an electrical circuit the electrician should always

(*a*) check that all his tools are readily available

(*b*) make sure that his life insurance premiums have been paid

(*c*) keep both hands in his pockets

(*d*) isolate the circuit and test to ensure that it is really dead

12M19 If a single-pole switch is used to control a circuit, it must always be connected in the

(*a*) upright position   (*b*) phase conductor

(*c*) brown wire   (*d*) neutral conductor

12M20 British Standard *BS 7671* is also known as the

(*a*) Electricity Supply Regulations

(*b*) Health & Safety at Work etc. Act

(*c*) IEE Wiring Regulations

(*d*) Electricity at Work Regulations

12M21 The resistance of the skin is normally high, but will be reduced, resulting in the possibility of higher shock currents, if

(*a*) extra-low voltage is used   (*b*) gloves are worn

(*c*) tools are not insulated   (*d*) it is wet

12M22 Probably the most effective form of artificial respiration to use on an electric shock victim is

(*a*) mouth-to-mouth resuscitation   (*b*) the Holger–Nielsen method

(*c*) heart massage   (*d*) the rocking stretcher

Chapter 13

Cables and enclosures

13.1 Supply system

The electricity supply in the United Kingdom is provided from generating stations which are sited close to a supply of fuel (such as the coalfields), close to cooling water (rivers or the sea) or close to heavy loads (large industrial areas). The name 'generating station' is technically incorrect. They do not *generate* electrical energy, but *convert* it from other forms, such as coal, oil, gas, nuclear, running water (hydro) and the wind.

Alternators operate at voltages in the range 11–32 kV. To reduce losses in transmission, it is necessary to use high voltages and transformers step up the voltage to 400 kV (the super-grid) or 275 kV for long-distance transmission. For shorter distances, voltage levels of 132 kV, 66 kV and 33 kV are used, and distribution to final substations takes place at 11 kV. For final distribution in rural areas where overhead lines are used, 6·6 kV or 3·3 kV supplies are sometimes used.

At the final substation, transformers step down the voltage to that used in the public supply which is 415 (400) V three-phase from which the 240 (230) V single-phase supply is derived. The secondaries of these transformers are almost always star-connected with the star point connected to earth.

13.2 Conductor materials and construction

Basic electrical conductors have already been considered in Section 1.4, showing that, from a purely electrical point of view, silver is the best conductor, but that its poor mechanical properties and high cost rule it out as a cable conductor material.

Copper

Copper is second only to silver as an electrical conductor, is easily drawn into wires, and is comparatively strong, but may be softened (annealed) to make it easier to bend. All small conductors, as well as many for heavy power cables, are made of copper.

245

Aluminium

Although aluminium is more than half as resistive again as copper, its density is less than one-third that of copper, so an aluminium conductor, although larger, will have half the weight of an equivalent copper conductor. The disadvantages are lower strength, higher thermal expansion, and rapid oxidisation which necessitates special jointing techniques. Aluminium conductors are almost always used for heavy-current overhead lines (often with a steel core for extra strength), and have been widely used for power cables. At the time of writing, aluminium cables of cross-sectional area less than 16 mm^2 are not allowed, although it is possible that this Regulation will be changed in due course. Table 13.1 shows the relationship between copper conductors and equivalent aluminium conductors.

Table 13.1: Standard sizes of copper cables with comparable aluminium sizes

Copper-conductor CSA (mm^2)	1·0	1·5	2·5	4	6	10	16	25	35
Aluminium conductor (of equivalent current rating) CSA (mm^2)	1·5*	2·5*	4*	6*	10*	16	25	30	50

*The use of these cables does comply with IEE Wiring Regulations (*BS 7671*).

Solid and stranded conductors

Some small cables, mineral-insulated and aluminium power cables have single-strand or solid conductors. The majority of cables, however, have stranded conductors to make them more flexible for installation purposes. The standard arrangements are for one, seven, 19 or 37 strands (Figure 13.1). Flexible cords and flexible cables have more numerous but smaller strands (Table 13.2).

It is possible to obtain standard wiring cables up to 2·5 mm^2 with multiple, as well as single, strands of conductor. Such cables are used in situations where their extra flexibility is useful.

When copper and aluminium conductors are drawn out into strands they become **hard-drawn**, in which condition they are stiff and hard. They can be **annealed** by heat treatment, when they become comparatively soft and pliable. Continuous bending of the conductors, particularly if the strands are large, will result in **work-hardening** and possible cracking.

Heating cables

Most electrical conductors are constructed so that they have low resistance. This results in small power losses and low temperature rises within the cable itself. Sometimes, however, cables are used for heating, and the conductors must then have a higher resistance. Such conductors are usually alloys containing some combination of copper, iron, steel, nickel and chromium. Heating cables are used for underfloor heating, keeping roads and ramps clear of snow, keeping gutters free of ice, soil heating for sports and horticultural purposes etc.

Figure 13.1   Conductor sections
  a Single strand
  b Seven strands
  c 19 strands

Table 13.2: Stranding of copper cables

Cross-sectional area (mm^2)	The first number indicates the number of strands; the second number gives the diameter of each strand in mm	
	Cables	Flexibles
0·5	—	16/·20
0·75	—	24/·20
1·0	1/1·13	32/·20
1·5	1/1·38	30/·25
2·5	1/1·78	50/·25
4	7/0·85	56/·30
6	7/1·04	84/·30
10	7/1·35	80/·40
16	7/1·70	126/·40
25	7/2·14	196/·40
35	19/1·53	276/·40

13.3   Cable insulators

A very wide range of cable-insulation material is now in use, some of the more important being listed with their properties below.

Polyvinyl chloride (PVC)

A basic plastics material which can be made in many forms. It is robust, chemically inert, has good aging and fire-resisting properties. It is more resistant to weathering and sunlight than rubber, but will harden and crack in the presence of oil or grease. High temperatures lead to softening and possible insulation failure, whereas at low temperatures, PVC may become brittle. Very little absorption of water will occur, although PVC is inferior to polychloroprene in this respect. PVC is widely in use as a sheathing material as well as for insulation. It can only usually be used where conductor temperature does not exceed 70°C. The material emits dense smoke and corrosive fumes when it burns.

Vulcanised rubber (VR)

A good electrical insulator with high flexibility. It is subject to rapid aging and cracking due to weathering and exposure to sunlight, and softens and becomes sticky in the presence of oils and greases. It ages rapidly at high or low temperatures. It is not widely used but is available in a form which will operate safely up to 85°C.

Polychloroprene (PCP)

Polychloroprene, also known as Neoprene, is a plastics material of lower strength and lower insulation resistance than PVC. However, it is more resistant than PVC to weathering, and to attack by oils, acids, solvents, alkalis and water. PCP is more elastic than PVC, and this elasticity is not affected by increased temperatures. Although more expensive than PVC, PCP is used as a cable insulation and sheathing material for conditions where PVC would not be suitable.

Silicone rubber

Silicone rubber is a synthetic material with many of the advantages of natural rubber. It has good weathering properties, and will resist attack from water and mineral oils, but not from petrol. It remains elastic over extremes of temperature ($-70°C$ to $150°C$).

Butyl rubber

Butyl rubber is another synthetic material which is less expensive than silicone rubber but has similar advantages. It remains flexible over the range $-40°C$ to $85°C$, but will burn readily once ignited.

Ethylene–propylene rubber (EPR)

EP rubber is generally similar to butyl rubber, but with improved properties. It is resistant to heat, water, oil and sunlight, and is suitable for direct burial, exposure to weather and contaminated atmospheres. It is used to insulate power cables which have an increased current rating owing to its heat resistance.

Chlorosulphonated polyethylene (CSP)

This material is mainly used for sheathing cables insulated with other plastics. It is capable of operating over the temperature range $-30°C$ to $85°C$, and has excellent oil resisting and flame-retarding properties. It is very tough, and very resistant to heat and water. This is the sheathing material which is called 'HOFR' (heat resisting, oil resisting and flame retarding) in the IEE Wiring Regulations.

Magnesium oxide

This material is in the form of a white powder, and is invariably used in mineral-insulated cables. It also finds application in some heating elements. Magnesium oxide is nonaging, and will not burn, but it is very hygroscopic (absorbs moisture from the air), when it loses its insulating properties. It is unaffected by high temperatures, and is a good conductor of heat.

Cross-linked polyethylene (XLPE)

This is a polmcric insulation (like EPR above) and is used particularly for power cables. In this application it shows cost advantages over oil-filled and pressurised cable insulation, not least in the simpler jointing techniques which may be used. The material does not burn easily and, when it does, only limited emission of smoke and fumes occurs.

Paper

Dry or oil-impregnated paper, wound in long strips over the conductors, was the most common method of insulating underground cables but has now been overtaken by XLPE. As long as the paper is kept dry (usually by sheathing with lead alloy) its insulating properties are excellent.

Glass fibre

Glass fibre, impregnated with high-temperature varnish, is used to insulate some flexible cords used for high-temperature lighting applications.

13.4   Bare conductors

Bare conductors are not covered with insulation. Those intended for use at voltages which are above a safe level will normally be supported on insulators. The IEE Wiring Regulations (*BS 7671*) approve the application of bare conductors where they are used as

(a) the external conductors of earthed concentric systems. This type of wiring uses an inner, insulated core for its phase conductor, and the outer conductor or sheath as a combined neutral and earth conductor

(b) the conductors of systems working at a safe voltage. An example is where galvanised-steel wire is buried in the soil and fed from a transformer at low voltage to provide soil heating

(c) protected rising-main and busbar systems. These systems use bare busbars enclosed within a steel trunking and supported on porcelain or plastics insulators. Provision is made for supplies to be tapped off at intervals. The system is often used as a mains-supply distributor for multistorey flats or for factories. See Figure 13.2.

Figure 13.2   Busbar trunking with plug-in type fused tapping box

(*d*) as current-collector wires for overhead cranes and the like. The conductors must then be well out of reach, and clearly labelled to indicate the danger.

13.5   Plastics- and rubber-insulated conductors

These cables are insulated, and, in some cases, may also have an overall braiding to protect them during installation. IEE Wiring Regulations (*BS 7671*) demand that they be protected against mechanical damage, and so they must be completely enclosed within a duct, trunking or conduit. Figure 13.3 shows typical constructions of insulated conductors.

Figure 13.3   Insulated conductors
        a PVC insulated
        b Elastomer insulated, braided and compounded

13.6   Sheathed wiring cables

For most domestic and commercial installations, the least expensive system includes the use of sheathed cables. These consist of conductors insulated with PVC laid up singly or with others, and covered by a protective sheath. The sheath may consist of PVC or PCP; PVC-sheathed cables are the most widely used, PCP finding particular application for outdoor installations (Figure 13.4).

Sheathed cables are fixed by saddles pinned to the surface, or may be buried in plaster walls, with or without the added protection of a conduit or capping provided that they run vertically or horizontally from an outlet such as a switch or a socket. Where run under floorboards, the cable must pass through holes which are at least 50 mm below the top of the joist to prevent damage by nails.

Joint boxes are used to connect sheathed cables, and, as with other terminations, the sheath must be continued right into the box, so that no unprotected conductors are exposed to damage.

A special sheathed cable with a further protective braiding and compounding is used on farm installations, where the corrosion problem is acute. Special joint boxes and accessories are available for this system.

Sheathed cables should not be manipulated or installed when the temperature is below 5°C, or there is a danger that the sheath or insulation may crack. If cables are tightly bent, the sheath may fracture; the minimum bending radius should never be less than three times the overall cable dimension and is larger for cables with conductors larger than 10 mm^2.

Figure 13.4   PVC-insulated PVC-sheathed cables
    a Single-core, stranded
    b Single-core, single-wire, with earth
    c Three-core, with earth, single-wire

13.7   Mineral-insulated (MI) cables

This cable consists of solid cores of high-conductivity copper embedded in highly compressed magnesium oxide, the whole being contained in a solid-drawn copper sheath. This sheath may be further protected by an overall covering of PVC, which will prevent deterioration in the presence of moisture and of most chemicals (Figure 13.5).

Standard cables are available with one, two, three, four or seven cores, with voltage ratings of 500 V and 750 V. Mineral-insulated cables are nonaging, have high mechanical strength, are small in size, are very resistant

Figure 13.5   Mineral insulated cable

a Components of two-core MI cable with PVC sheath overall
b Standard conductor arrangements

to corrosion, have excellent electrical properties, are waterproof, can work at high temperatures, and are suitable for use in flameproof installations. They must be terminated with special seals (Section 13.9) to prevent the insulator becoming moist, when it will lose its insulating properties.

The hard-drawn sheath may crack owing to work-hardening if bent too often, or if bent too tightly. Usually the minimum bending radius must not be less than six times the cable diameter, although if the cable bent only once, the minimum radius may be three times the cable diameter.

There is often no identification of the cores of a multicore MI cable, which must be checked through with a continuity tester and marked with coloured sleeves.

The copper sheath of this cable must be kept separate from the metalwork of water and gas systems or, where this is not possible, must be securely bonded to them.

Mineral-insulated cables with aluminium conductors and sheath are also available.

13.8   Armoured cables

An armoured cable is one where the conductors and insulation are protected by a layer or layers of steel wires or tapes. This armouring protects the cable from mechanical damage while it is in service, as well as giving added strength to withstand handling during installation.

The conductors of these cables are made of copper or aluminium, the former always stranded, but the latter often solid. Conductors are often sector-shaped (Figure 13.6) to fit together more closely.

Dry or impregnated paper has been in use for many years as the most usual insulator for these cables, with a lead sheath to keep out moisture. In recent years, PVC has been used for both insulation and sheath of lower-voltage cables, with XLPE as insulation for both low- and high-voltage types. A soft bedding of jute or other suitable material is sometimes applied over the sheath to serve as a base for the steel armour.

Figure 13.6   Armoured cable with thermosetting insulation

Steel or aluminium wires used as armouring are given a twist, or 'lay', along the cable so that they fit neatly into the bedding. Steel-tape armouring may be wound round the cable in a spiral, or may be in the form of longitudinal strips. The armouring is usually, but not always, protected from corrosion by an overall 'serving' of impregnated jute, or of PVC.

Power cables of these types find wide application for underground electrical supplies, as well as for the heavy submain distributors in some large buildings. They are also used for circuit wiring in factories where the likelihood of mechanical damage is great.

Paper-insulated cables are slightly more expensive, and are more difficult to joint (Section 13.9) than the PVC-insulated and XLPE-insulated types, but are able to carry more current for a given size, because increased conductor temperature does not affect the paper insulation. Figures 13.6, 13.7 and 13.8 show some examples of armoured cables.

Figure 13.7   PVC -insulated PVC-sheathed steel-wire-armoured cable

Figure 13.8   PVC-insulated PVC-sheathed steel-wire-armoured cable with solid-aluminium conductors

13.9   Cable joints and terminations

Cables must be terminated by connecting them at control gear, accessories etc., but joints along the run of a cable should be avoided wherever possible.

Terminations and joints must be electrically and mechanically sound, the principles being that electrical resistance should be low, and insulation resistance high.

Cables of differing types require different treatments, terminations and joints, some of the most common cables being detailed below.

Rubber- and plastics-insulated cables

There is no need with these cables to seal the cable ends against the ingress of moisture, and the pinching screw is the most common type of terminal (Figure 13.9). Some types of single-strand conductor, particularly of aluminium, have been found to work loose in time owing to deformation of the conductor material. To prevent this difficulty, the pinch screw should not be overtightened; the pressure-plate type of terminal gives good results with single or stranded cores.

Where larger cables must be terminated, for example in switchgear, the grip type of cable clamp may be employed, gripping the whole of the conductor and not causing damage.

It is still fairly common for large cables to be connected to switchgear or busbars by fixing lugs to the conductor ends. The traditional method is to sweat the lug solidly to the conductor using tin-man's solder and a noncorrosive flux (Figure 13.10). An alternative method, known as 'crimping', is to fix the lug to the cable by means of pressure and indentation. The conductor is inserted into the lug, and an indentation made by using a hand tool or a hydraulic press. The pressure compacts the lug and the

Figure 13.9   Screw terminals
a Well designed pinch-screw-type terminal
b Pressure-plate-type terminal

Figure 13.10   Socket soldered to cable
Insulation removed for soldering or damaged by heat is made good with tape

Figure 13.11   Pressure lug cut through to show effect of 'cold welding'

cable into a single solid mass, so that the joint is stronger than the cable itself (Figure 13.11). This type of termination is quicker and cheaper than the sweating method, and causes no damage to the insulation. The compression lug is used with both copper and aluminium cables, whether stranded or solid. It is particularly useful with aluminium cables, which are difficult to solder owing to the very rapid oxidisation of surfaces after cleaning.

Aluminium may corrode when in contact with other metals, especially in the presence of moisture, or may cause corrosion of the other metal. Aluminium-to-aluminium joints should always be made when possible, but, where this cannot be done, special precautions may need to be taken. These include electroplating of the other metal, or coating it with a layer of grease or compound before connection. Corrosion is particularly likely when aluminium is in contact with a metal having a high copper content such as brass.

Mineral-insulated cables

The mineral insulation of these cables is very hygroscopic. This means that the insulation will absorb moisture from the air. Wet insulation of this type loses its insulating properties, so the cable ends must be sealed to prevent moisture getting in.

The sealing pots used are screwed or wedged on to the cable sheath, and enclose the moisture-resisting compound, which is compressed by crimping a disc into position in the pot end. The disc also secures the PVC sleeves used to cover the bared copper conductors. The completed seal is either enclosed in a gland, an 'exploded' view of one type being shown in Figure 13.12, or can be accommodated in a 'glandless box' (Figure 13.13). The conductors are terminated in pinch- or clamp-type terminals, or by means of cone grip, sweated or pressure lugs.

Flexible cords and cables

Flexible cords and cables are usually connected to portable apparatus and plugs, the normal method being by wrapping the conductor round a captive screw under a washer. If the conductor is wrapped in a clockwise direction,

Figure 13.12   Terminating mineral-insulated cable
a Seal
b Gland

Figure 13.13   Glandless boxes for use with MI cables

it will tighten, and not push out as the screw is tightened. Pinch-type terminals are used in some cases, but extra care is needed in the manufacture, as any sharpness on the end of the pinching screw may sever the small conductor strands.

Lugs of both sweated and pressure types can be used with flexibles. With sweated lugs, special care is needed; if the heat is applied for too long, solder will creep up the conductor between the strands. If it does so, that section of the conductor will become rigid, and a break will occur in due course owing to bending at the junction of the flexible and rigid sections of the conductor.

Armoured cables

Armoured cables are terminated or joined with the aid of special glands or sealing boxes. PVC-insulated and PVC-sheathed types use glands, of the type shown in Figure 13.14. The sealing ring shown over the sheath at the end of the gland may be omitted if the gland is for use indoors, because the PVC insulation is not hygroscopic.

Paper-insulated cables have a hygroscopic insulation which must be sealed. Joints are made by the compression method or by sweating the conductors in copper ferrules, the insulation being made good with tape. The joint or termination is then enclosed in a box, which is filled with a moisture-proof insulating compound.

Figure 13.14   Gland for use with PVC-sheathed armoured cable

13.10   Conduits

If, as in most electrical installations, it is necessary for a number of conductors to follow the same route, it is often convenient and economical to use insulated-only conductors and to enclose them in a common protective tube. Such a tube is a conduit, and is most usually made of steel, although plastics, aluminium and copper conduits are also available. The conduit system is erected complete before the cables are pushed or drawn into it, inspection boxes often being used if long runs or numerous bends are involved.

Cable capacities of conduit and trunking

The number of cables of a given size which can be fed into a certain conduit will obviously be limited by physical considerations, but will be further reduced by electrical requirements. For example, if cables are tightly packed, insulation will be compressed, its thickness reduced and its effectiveness impaired. More importantly, cables become warm in service owing to the current they carry. If this heat cannot escape, and too many cables in a conduit make heat transfer more difficult, the cables will overheat, and insulation may be damaged.

To prevent overcrowding of conduits, IEE Wiring Regulations: Guidance Note No. 1, Appendix A provides tables for common cable, conduit and trunking sizes.

Light-gauge-steel conduit

These conduits are often used in straight lengths with rubber-bushed ends for the protection of sheathed cables where these are buried in plaster. They are made of thin-section material, often no thicker than 1 mm, which is bent into a circular or oval shape with the edges butted, and often not joined together as in heavy-gauge conduit. Circular light-gauge conduit may be erected into a complete system by the use of grip-type boxes, but, since it cannot be bent successfully, its use is very limited.

Heavy-gauge-steel conduit

As the name implies, this conduit is thicker-walled than the light-gauge types, with wall-thicknesses of 1·65 mm and above. The conduit is erected by threading it, some conduit connections being shown in Figure 13.15. Joints are accomplished by means of couplers (sockets) (Figure 13.15*a*), with circular or rectangular boxes providing for drawing-in of cables or mounting accessories. Connections to boxes may be by means of threaded spouts provided, or using a male bush and socket or a female bush to connect to a clearance hole (Figure 13.16).

In most industrial applications, there is no need to hide conduit, and it is fixed to the surface of walls and ceilings by means of saddles. To reduce the resistance of the conduit system when used as a protective conductor, to allow for easy drawing-in of cables and to prevent damage to the insulation, conduit ends must be butted up in couplers, and all burrs must be removed.

Conduits are often chosen to accommodate cables for installations in concrete buildings, such as flats and office blocks, and here the installation

a   b

Figure 13.15   Conduit connections
a Conduits tightly butted in socket
b Running socket with locknut

Figure 13.16 Conduits connected to clearance hole
  a Male-bush and socket method
  b Female-bush and locknut method

must be hidden. This is frequently accomplished by burying the conduit in the concrete itself, special outlet boxes being available for this purpose (Figure 13.17). Heavy-gauge conduit is bent where necessary using a bending machine; the bending radius must never be less than $2 \cdot 5$ times the outside diameter of the conduit. Where a long vertical run of conduit is involved, cables must be supported within the conduit at intervals of not more than 3 m to prevent strain. Fire barriers must be placed within a conduit where it may otherwise allow the spread of fire. Heavy-gauge-steel conduit usually forms the protective conductor.

Figure 13.17 Standard and angle-pattern looping-in boxes

Plastics conduit

Plastics conduits, usually of hard PVC, are widely available as an alternative to steel conduits. They cannot, of course, be used as circuit protective conductors as can steel conduits, but their lightness, cheapness and speedy fixing more than repay the need for an additional protective conductor. They do not have the same mechanical strength as steel conduits, become soft above 60°C and brittle below − 15°C, but are increasingly used for both buried and surface work.

   Joining is sometimes by threading, but more usually by a pushfit taper joint, often used in conjunction with an adhesive.

Nonferrous-metal conduits

Aluminium conduits are available, with sherardised threaded inserts suitable for use with the normal steel-screwed-conduit fittings. Copper and zinc-base-alloy or stainless steel conduits are occasionally used in special situations.

13.11   Ducts and trunking

Ducts

A duct is a closed passageway formed within the structure of a building into which cables may be drawn. It may consist of a steel or plastics trunking buried in the floor or ceiling of a building, or may be cast *in situ*. This means that a former is placed before concrete is poured; when it is removed after the concrete has set, it leaves the duct.

If all or part of a duct is of concrete, only sheathed cables must be drawn into it.

Trunking systems

A trunking is a rectangular-section enclosure, most usually made of sheet steel, but also available in plastics. Where large numbers of cables follow the same route, a trunking system will be used instead of a very large conduit, or a multiple run of smaller conduits, because it is lighter, cheaper, quicker and easier to install. Cables can be laid into the trunking before the lids are closed, instead of being drawn in as with conduits and ducts. Special fittings are available for junctions and bends (Figure 13.18).

As well as its use for longer runs of cabling, trunking can be installed with advantage to enclose the interconnecting and circuit cables at mains-switchgear positions. In some cases, it is necessary to keep circuits separated. This is often necessary where telephone and fire-alarm cables follow the same route as the supply cables; multicompartment trunking can then be used (Figure 13.19).

13.12   Cable ratings

The rating of a cable is the amount of current it can be allowed to carry continuously without deterioration. The rating depends entirely on the temperature the cable is capable of withstanding without deterioration of its insulation, although, with unsheathed mineral-insulated cables, the insulation of which will not be impaired no matter how hot it gets, the effect of high temperatures on the seals and on persons touching the sheath or the possibility of burning the surface to which it is fixed are the deciding factors.

Figure 13.18   Cable trunking and fittings

Many factors govern the rating of a cable. These are

(a) *conductor cross-sectional area:* the larger the cable, the more current it will carry without getting too hot

(b) *insulation:* some types of insulation will be damaged at temperatures where no deterioration occurs with others; for instance, a PVC insulated $2 \cdot 5$ mm^2 cable may be rated at 24 A, whereas a mineral-insulated conductor of the same size may be rated at 43 A

(c) *ambient temperature:* if a cable is installed in a hot situation, it will be unable to dissipate the heat it produces as quickly as it would in cold surroundings, and will operate at a higher temperature

a

b

Figure 13.19   Special trunking
a Multiple-compartment-type
b Skirting-type

(*d*) *type of protection:* in the event of a fault or of an overload, the speed with which the fuse blows or the circuit breaker opens may differ very widely; if a fault current flows, the temperature of the cable carrying it will increase with time, so the quicker the protective device operates, the less cable heating will occur. Thus, cables protected by rewirable fuses are rated at lower currents than those protected by HBC fuses or circuit breakers

(*e*) *grouping:* if a large number of cables is run together, say in a conduit or trunking, they will be unable to radiate and conduct heat as freely as if they were installed separately. Cables which are in larger groups are hence rated lower than those run singly

(*f*) *disposition:* a cable buried direct in moist earth will conduct heat to its surroundings more rapidly than a cable lying in a duct, and can be rated at a higher current

(*g*) *type of sheath:* if a cable is armoured with a second overall sheath, it is more effectively heat-insulated than one without armour, and must be given a lower rating if the insulation has the same properties

(*h*) *contact with thermal insulation:* a cable in contact with, or buried in, thermal insulation, will be prevented from dissipating heat normally. This situation will often arise in roof spaces, filled cavities of walls and so on.

The electrical craftsman may have difficulty in deciding what weight to give to each of the above factors. The IEE Wiring Regulations (*BS 7671*) provide detailed current-rating tables for all types of cable, together with rating factors to allow for such factors as grouping and disposition. An understanding of these tables is essential for all who must choose a cable for a particular installation.

13.13   Exercises

1 List the materials used as conductors for power cables, and compare their advantages.

2 Give the names of *four* materials used to insulate cables, and state where each could be used with advantage.

3 The following materials are used in the manufacture of cables. By giving the important properties of each, show where it could be used.
   (*a*) copper
   (*b*) aluminium
   (*c*) PVC
   (*d*) PCP
   (*e*) paper
   (*f*) glass fibre.

4 List *three* situations in which bare conductors may be used.

5 Describe the construction of a twin-with-protective-conductor sheathed wiring cable. What are the advantages of this type of cable, and in what type of installation is it likely to be used?

6 Using a sketch, describe the construction of a mineral-insulated cable. Show how this cable is terminated, and explain why a special termination is necessary.

7 List the advantages of mineral-insulated cables. Give two examples of installations where this type of cable would be used.

8 Describe the construction of a PVC-insulated PVC-sheathed steel-tape-armoured cable.

9 Describe the compression (crimping) method of fixing a lug to a conductor, and compare with the sweating method.

10 What precautions must be taken when connecting an aluminium conductor to a brass clamp? Why are these precautions necessary?

11 Use sketches to show how
   (*a*) two conduits are connected in a coupler
   (*b*) a conduit is connected in a spout box
   (*c*) a conduit is connected to a clearance hole.

12 What are the advantages of cable trunking compared with steel conduit?

13 List the factors affecting the rating of a cable. Explain why each factor affects the current the cable may carry continuously.

13.14   Multiple-choice exercises

13M1   The highest voltage used for transmission of electrical energy in the UK is
   (*a*) 375 kV   (*b*) 400 kV   (*c*) 415 V   (*d*) 11 000 V

13M2   The most usual conductor material for cables is
    (*a*) aluminium   (*b*) sodium
    (*c*) silver   (*d*) copper

13M3   Aluminium is often used as conductor material on heavy power cables rather than copper because
    (*a*) it is cheaper and lighter
    (*b*) it is likely to corrode more easily
    (*c*) it will work harden and break when repeatedly bent
    (*d*) it can be safely fixed with copper or steel saddles

13M4   A flexible cord or cable will have a conductor made up of more strands than a fixed wiring cable of the same cross-sectional area because
    (*a*) it is easier to make it that way
    (*b*) if soldered, the conductor may lose its flexibility
    (*c*) the strands can be twisted together before being connected
    (*d*) it will bend more easily

13M5   A disadvantage of the common insulation material PVC is that
    (*a*) it is extremely expensive
    (*b*) it has a complicated name
    (*c*) dense smoke and corrosive fumes are emitted when it burns
    (*d*) it can be extremely difficult to strip

13M6   A cable insulation which has the properties that it is resistant to heat, water, oil and sunlight, may be buried directly in the ground and exposed to weather is
    (*a*) ethylene–propylene rubber (EPR)
    (*b*) polychloroprene (PCP)
    (*c*) magnesium oxide
    (*d*) cross-linked polyethylene (XLPE)

13M7   The insulating material used in mineral-insulated cables is
    (*a*) PVC   (*b*) magnesium oxide
    (*c*) sulphur dioxide   (*d*) hard-drawn copper

13M8   An armoured cable is likely to be chosen for installation
    (*a*) in the kitchen of a normal house
    (*b*) overhead between a house and a shed
    (*c*) of a temporary nature which will be replaced within three months
    (*d*) in a heavy industrial situation where there is a danger of mechanical damage

13M9   An insulator which absorbs moisture from the air is called
    (*a*) hygroscopic   (*b*) cross-linked polyethylene
    (*c*) absorbent   (*d*) wet

13M10 One reason for limiting the number of cables drawn into a conduit is
    (*a*) so that the cables are in very close contact with each other
    (*b*) to limit the cost of the installation
    (*c*) to ensure that cables can be easily installed and removed
    (*d*) so that the conduit does not become crowded

13M11 Heavy-gauge steel conduits are joined together by
    (*a*) push-fit connectors and adhesive
    (*b*) welding or brazing
    (*c*) threading and threaded couplers
    (*d*) butting firmly together

13M12 Where large numbers of cables are concerned, the use of trunking rather than conduit is an advantage because
    (*a*) trunking is rectangular in section rather than circular
    (*b*) cables can be laid in rather than having to be drawn in
    (*c*) the cables can be identified more easily
    (*d*) bends can be tighter in trunking than in conduit

13M13 Different categories of circuit can be run in a common trunking if
    (*a*) they are separated from each other by compartments
    (*b*) they are each insulated for the voltage they carry
    (*c*) they are taped together
    (*d*) they are separated within the trunking by at least 10 mm

13M14 A cable buried in thermal insulation must be derated because
    (*a*) it will be difficult to find if a fault develops
    (*b*) the thermal insulation will attack the cable sheathing
    (*c*) the thermal insulation may be inserted after the electrical installation is complete
    (*d*) it will be less able than others to shed heat and will become hotter

13M15 A cable protected by a semi-enclosed fuse must be derated because
    (*a*) the fuse may take longer to break circuit current in the event of a fault or an overload
    (*b*) semi-enclosed fuses are cheaper than other forms of protection
    (*c*) the fuse will be faster acting than the alternatives
    (*d*) the wrong size fuse wire may be fitted

Chapter 14
Lighting and heating installations

14.1 Introduction

Having considered the principles of electrical technology, the precautions necessary for safety, and the cables which can be used for installations, we now have to consider the electrical installations themselves. Electrical installations can vary very widely in scope and in complexity, from, for example, a single lighting point on the one hand, to the complete installation of an oil refinery on the other. All installations follow the same basic principles to ensure safety from fire and shock. However, the mains and distribution parts of large industrial systems are complicated, and will not be considered here; domestic and small commercial installations form the majority of the total of electrical installations. Only this type of work is considered in the following sections.

14.2 Supply-mains equipment

The supply-mains equipment is that normally fixed close to the point where the supply cable enters a building. This equipment is designed to protect the rest of the installation against excessive current, and falls into two categories: that which is the responsibility of the supply company, and that provided and maintained by the consumer.

Supply company's mains equipment

The incoming supply cable must be terminated, usually in a sealing box, and the mains cable then feeds to a main fuse or fuses. The main fuse, or service fuse, almost always of the HBC type, protects the feeder cable against short circuits at the mains position, and against overloads and short-circuits in the consumer's installation. In a properly planned and maintained system, the consumer's fuses or circuit breakers will operate before the main fuse blows; but this fuse is also the insurance by the supply company against a badly maintained installation. For example, if a consumer's 30 A ring-main fuse blows owing to a fault, and the consumer increases the fuse rating in an attempt to keep the circuit operating, the supply company fuse will blow to protect the supply cable.

From the main fuse, the supply feeds on to the meter, which records the energy used so that the consumer can be charged. The sequence of supply equipment is shown in Figure 14.1.

All of the supply company's equipment is sealed because, if connection is made to the supply *before* the meter, the consumer cannot be charged for energy used.

Consumer's supply equipment

The consumer's supply equipment must control and protect the installation against overloads and faults. Equipment is connected to the supply company's meter in the following sequence:

(a) *main switch*: which must break both poles of the supply: this means that for the usual single-phase AC installation the switch must be of the double-pole linked type

(b) *main fuse:* often this main fuse can be omitted, the supply company's fuse giving protection

(c) *distribution board:* this board has fuses or circuit breakers to protect the outgoing final circuits.

Figure 14.1   Layout of supply company's equipment

In most new domestic installations, the supply equipment comprises a consumer's control unit. This unit contains a 60 or 100 A double-pole main switch, as well as up to eight fuses or circuit breakers of 45 A, 30 A, 15 A or 5 A rating. The fuse or circuit-breaker rating must not be more than the current rating of the smallest conductor it protects; thus 5 A fuses are used for lighting circuits, 15 A fuses for immersion heaters and similar loads, 30 A fuses for cookers and ring-main socket circuits and 45 A for the larger cookers (Figure 14.2).

Increasing use is being made of residual-current devices (RCDs) to protect against electric shock due to indirect contact and against fire. Although it is quite possible that the whole installation may be protected by the connection of an RCD immediately before the double-pole main switch, this could cause problems in that the complete installation will be switched

a

b

Figure 14.2   Consumer's service units
a Schematic of eight-way unit
b Six-way unit

off when the RCD operates; for example, there may be danger if the lighting is lost as a result of an earth fault in an appliance connected to a ring circuit.

A more usual solution would be to install a split consumer's unit, which has some of its circuits protected by an internal RCD. A possible arrangement is shown in Figure 14.3.

In some cases, one or more additional RCDs may be required, for example to give protection to the user of an electric lawnmower by providing a socket outlet protected by its own RCD. It is important in such cases that the RCD in the consumer's unit should be time delayed. This prevents it operating before the dedicated RCD protecting the circuit with the fault which would result in the loss of more circuits than necessary.

In larger domestic or commercial installations, more than eight circuits will be necessary, and the supply equipment must be larger. There are many possible variations, but a common layout is shown in diagrammatic form in Figure 14.4.

The main switchfuse or fuse switch consists of a linked switch and fuses enclosed in one case. The busbar chamber encloses bare copper or aluminium bars mounted on insulating supports, and is used as a connecting box to couple together the output of the main fuse switch, and the main circuits.

The main circuits are controlled and protected by switchfuses mounted close to the busbar chamber. These main circuits may feed out to individual heavy loads, or may feed distribution fuseboards from which lighting and other final circuits are fed.

Figure 14.3 Split consumer's unit with RCD

Figure 14.4   Layout of supply equipment for office block

14.3   Lighting circuits

It is not good practice for a lighting circuit to feed a total load exceeding 15 A. This means up to 36 lights connected to one final subcircuit. In the normal installation, good planning usually limits the number of lights on each circuit to about ten, with more than one lighting circuit to each building. This ensures that the whole of a building is unlikely to be plunged into darkness by the operation of one fuse.

Simple theoretical circuits for the control of lighting points are shown in Figure 14.5. In practice, the connections used will depend on the wiring system chosen. The loop-in method (Figure 14.6) is used where cables run through conduit, all connections being made at switches or lighting points. Other joints may be difficult to find in the event of a fault. The conduit can be used to provide the earthing point needed at each outlet if it is of the heavy-gauge-steel type; if light-gauge or plastics, an extra PVC-insulated protective conductor (coloured green and yellow) must be used.

Where a sheathed wiring system is used, an included protective conductor will be connected to the earthing points. Single-core sheathed conductors, with the addition of a protective conductor, may be used, connections being

Figure 14.5 Basic lighting circuits
a Light controlled by one-way switch
b Light controlled by two two-way switches
c Light controlled by two two-way switches and one intermediate switch.
Light is controlled from any one of three switches.
Alternative switch connections are shown dotted.

Figure 14.6 'Loop-in' wiring system
Earth connections at outlets are made to metal conduit system

the same as for the loop-in system. It is far more common, however, for joint boxes to be used. These boxes may have fixed terminals, or may enclose loose connectors, and are securely fixed to joists under floors or in ceiling spaces. Where situated under floors, they must be covered by a screwed trap in the floor, so that they can be examined if necessary. Connections for the joint-box system are shown in Figure 14.7.

An alternative system for use with sheathed cables requires a special ceiling rose with an extra terminal, which must be shielded so that it cannot be touched when the cover is removed for replacement of the flexible cord. This precaution is necessary since the third terminal of these 'three-plate'

Figure 14.7   Joint-box wiring system

ceiling roses is used for connecting phase conductors, and remains live when the controlling switch is off. Connections are shown in Figure 14.8.

It is sometimes necessary for an existing light controlled by a one-way switch to be modified for two-way switch control. Figure 14.9 shows two methods of feeding the extra switch.

Figure 14.8   Three-plate ceiling-rose wiring system

14.4   Socket-outlet circuits

There are two basic types of socket outlet, and these will be considered separately, since the circuits are quite different.

Sockets for unfused plugs

These sockets accept round-pin plugs which, although available with fuses in the plug, are usually unfused. They are available in 15 A, 5 A and 2 A

Figure 14.9 **Alternative methods of converting existing one-way switch to two-way**

The first of the two-way conversions shown should be used where an induction-loop system for the deaf is in installed, because it causes much less interference than the other methods

ratings. Each 15 A socket must be wired back to its own circuit fuse (Figure 14.10*a*). 5 A sockets may be looped on a radial final circuit, with not more than three sockets on each circuit (Figure 14.10*b*). For three 5 A sockets on one circuit, a 15 A fuse is necessary, so the current rating of the cable used must be 15 A throughout. 2 A sockets are normally used to feed portable lighting fittings, and can conveniently be connected to lighting circuits. Although sockets of this type are still in use, they must now be considered obsolete.

Sockets for fused plugs

These sockets will only accept plugs which are fused and use the flat-pin type complying with *BS 1363*.

Figure 14.10 **Typical circuits for sockets with unfused plugs**

The fuse in the plug protects the flexible cord and the appliance, so the circuit fuse now protects only the circuit cable and the socket (Figure 14.11). The system thus has two advantages:

(*a*) Since the plug is connected directly to the appliance, the fuse in the plug (13 A or 3 A) can be chosen to give the best protection. For example, a table lamp would be protected by a 3 A fuse, whereas a 2 kW fire would have a 13 A fuse fitted in its plug.

(*b*) The number of sockets connected on one circuit can be related to the *total* demand required, without having to assume that each socket always provides its full rated output. Thus, say, six 13 A sockets may be connected to a circuit protected by a 20 A fuse. If each socket were fully loaded, total circuit current would be 6 × 13 A = 78 A, and the circuit fuse would blow but, if all sockets are in the same room, this is very unlikely to happen. We say that **diversity** can be applied to the circuit.

Figure 14.11 Role of fuses in ring and radial circuit

Figure 14.12 shows the various circuit arrangements which are permissible with fused-plug sockets, and applied to domestic situations only. A fixed appliance, such as a small water heater, may be connected through a fused spur box, but only one such outlet may be fed from a spur.

The use of fused-plug circuits, especially of the ring circuit (Figure 14.12*c*), has permitted the provision of adequate sockets at reasonable cost.

14.5   Other circuits

The majority of the electrical apparatus in a modern home or office is connected to the lighting circuits or is fed from sockets. There are, however, equipments where this is either impossible or undesirable. For example, a large electric cooker must be fed back on a separate circuit connected to its own separate fuse or circuit breaker, usually rated at 30 A or 45 A. Space-heating appliances and immersion heaters fitted to tanks with a greater capacity than 15 litres must be fed from suitable separate circuits.

Figure 14.12   Final circuits for sockets with fused plugs

All connections are three-core, with phase, neutral and earth connectors. The number of sockets on the circuit is unlimited by the floor area served and must not exceed
a 20 m²   *b* 50 m²

c All connections are three-core, with phase, neutral and earth conductors. The number of fused spurs is unlimited. Each unfused spur must not feed more than two single or one twin socket, and the number of sockets and fixed appliances fed from unfused spurs must not exceed the number connected directly to the ring circuit.

At the other end of the scale, most houses have an electric bell. The circuit is fed at low voltage from a transformer, usually situated at the consumer's service-unit position. As well as being safe, the wiring can then be carried out using small, easily-hidden and inexpensive bell wire. A typical circuit is shown in Figure 14.13.

Figure 14.13   Typical domestic-bell circuit

Alarm systems

Losses owing to fires and break-ins have increased year by year, and the demand for fire and intruder alarms has increased with them. A simple open-circuit alarm system which may be used for fire or intruder purposes is shown in Figure 14.14. Closing any switch will sound the alarm.

The form of the switches will vary depending on the application. For fire alarms they will consist of 'break glass' points, where a spring-loaded plunger moves out to close contacts when a thin sheet of glass which normally holds it back is broken; or an automatic system will include heat or smoke detectors which are arranged to close on operation. Recent developments in automatic operation in the event of a fire or of a burglary include passive infra-red detectors (which sense the temperature increase due to a fire or to an intruder), ultrasonic and sonic detectors, and microwave detectors. With intruder alarms, it is more difficult to arrange for a switch operated by the unwelcome person to remain closed, and a continuous-ringing bell (or a relay to convert a normal bell into one) may be necessary (see Figure 7.3). A battery, often trickle charged from a mains-fed DC supply is preferred to a transformer operating directly from the public supply because of the fear of failure in the event of fire, or of interference by the intruder. The same fears will be felt for the circuitry itself; a cut wire or a burned cable could put the system out of action.

Figure 14.14   Simple open-circuit alarm system

For this reason, **closed-circuit** alarm systems are usually preferred. Instead of contacts which must close to operate the circuit, normally-closed circuit switches are wired in series as shown in Figure 14.15. The relay contacts which operate the bell are held open by a coil energised through the alarm contacts. A steady small current in this series circuit monitors its condition. In the event of a contact opening, or a deliberate or accidental break in a circuit, the relay will drop and sound the alarm. A pair of auxiliary contacts may be used in conjunction with a telephone line to alert the police or fire service.

Figure 14.15   Simple closed-circuit alarm system

This system has much to commend it, because it 'fails to safe'. Fine wire can be stretched across windows and other likely entry points so that it is likely to be broken by illegal entry. Many sophisticated alarms activated by heat, smoke and flames for fire systems or by body heat, movement, or sound for intruder alarms are not uncommon.

To reduce the risk of fire-alarm cables being burned through before they can operate, all such cables must be segregated or separated by fireproof barriers from all other cables, unless they consist of mineral-insulated cables, which are unaffected by high temperatures.

14.6   Earthing and polarity

The principles of earthing were considered in Chapter 12. Non-current-carrying metalwork must either be protected so that it cannot be touched in the event of a voltage to earth, or connected to earth so that no potential difference can exist between the metalwork and earth. For electrical installations, the connection to earth is made by joining all such metalwork together by means of the protective conductor, which is itself then connected to earth.

The protective conductor may take a number of forms, such as a separate insulated conductor, the earthing conductor in a sheathed cable, the earthing conductor in a flexible cord (green-yellow core), the sheath and/or armour of a cable, steel conduit, trunking etc. To ensure that lighting switches, lighting fittings and sockets can be earthed effectively, an earthing terminal must be provided in the outlet box serving them (Figures 14.7 and 14.8). This is necessary even though the switches or fittings are all-insulated; should they at some future time be changed for metal types, earthing could be effectively carried out. The earth connection of a socket must be coupled to the earthing terminal in the socket box by a short earth connector. The connection between the box and the socket by means of the fixing screws is not relied on for earthing purposes.

The importance of polarity has been stressed in Chapter 12. Where single-pole switches are used, care should be taken to ensure that they are connected in the phase (unearthed) conductor (Figures 14.6, 14.7 and 14.8).

Figure 14.16   Connections for testing of protective conductors
All fuses out and circuit breakers off
Main switch off
All sockets to be tested

wander lead

Figure 14.17   Measurement of system resistance with wander lead
In this case the resistance of the protective conductor to be
measured will be $0 \cdot 26 - 0 \cdot 15 = 0 \cdot 11 \ \Omega$

14.7   Simple testing

All electrical work must be inspected and tested on completion, and this
applies particularly to electrical installations. Part 7 of the IEE Wiring
Regulations (*BS 7671*) is devoted to inspection and testing, as is IEE
Guidance Note 3. Whilst inspection will reveal obvious deficiencies, such
as the failure to secure a cable, testing will be necessary to identify faults
such as wrong connections, poor insulation and so on.

Continuity tests

The objective of continuity testing is to ensure that conductors are
continuous and unbroken along their entire length. A simple continuity
test can be carried out using a bell or buzzer with a battery to pass a low
current through the cable under test, but it is very much more likely that
a dedicated resistance-measuring instrument will be used. Such devices
often can also be used for insulation resistance measurement.

Actual resistance readings for the protective conductor will be necessary
if a circuit design relies on protection from electric shock by limitation of

the shock voltage received in the event of a fault (Regulation 413–02–12). These readings can be taken with an instrument such as that mentioned above as indicated in Figure 14.16; a temporary link is required between phase and earth at the supply position, and of course this link *must* be removed after testing. Allowance must be made for the resistance of the phase conductors included in the measured value.

An alternative is to use a long (wander) lead to connect the measuring instrument directly across the conductor system whose resistance is to be measured. This will involve measurement of the wander-lead resistance, and its subtraction from the total measured value to give an accurate reading for the conductor system. Many modern instruments have a facility for storage of the measured lead resistance and its automatic subtraction. Figure 14.17 shows the arrangement.

The continuity of ring circuits must also be tested to ensure
(i)   that rings are actually continuous and
(ii) that there are no 'bridges' across the ring.
The tests are somewhat complicated and full details of the steps to be taken will be found in Section 8.2 of IEE Guidance Note 3.

Insulation-resistance tests

The insulation resistance of a system is the resistance across the insulation, both between phase and neutral and between phase and earth. In a sound installation with effective insulation, the value will be high, but dampness, incorrect connections or poorly made and insulated joints may reduce it to a lower value.

The test must be carried out with a direct voltage as the use of an alternating voltage will give a faulty reading because the insulation is behaving as a capacitor. The value of the test voltage will depend on the voltage of the circuit under test, and will be 500 V for a standard installation. This test voltage may be produced by operation of a hand generator but is more likely to be generated electronically from batteries in the test instrument. It is most important that the test instrument used will maintain the required test voltage when the current it supplies reaches a value of 1 mA.

The high voltage used for testing will damage some items commonly found in an electrical installation, and they must be disconnected before testing. Such items include dimmer switches, touch switches, delay timers, power controllers, electronic starters for fluorescent lamps and passive-infra-red detectors (PIRs) used in fire and intruder alarms. Such items must be separately tested at a suitable lower voltage before reconnection, when their insulation-resistance values should not be less than the value given in the appropriate British Standard, or $0 \cdot 5$ MΩ where there is no such standard.

The Regulations indicate that the minimum acceptable insulation resistance is $0 \cdot 5$ MΩ, but this value is very low for a healthy installation unless it is a very large one with many insulation-resistance values in parallel to give a low overall reading. If the measured value is less than 2 MΩ on a small or medium-sized installation further investigation will be necessary to find the reason for the low reading.

Tests should be carried out with the main switch off, all fuses in place (or circuit breakers closed), lamps removed and other equipment disconnected. Where lamps cannot be removed, their local controlling switches should be open. The test should be carried out:

(i)   between phase and neutral (P–N)
(ii)   between earth and phase and neutral connected together (PN–E). The connections for such a test are shown in Figure 14.18. Care must be taken to ensure that the temporary link from phase to neutral inserted for the second test is removed when the test is completed.

Figure 14.18   Insulation-resistance tests
All switches closed (including two-way)
All lamps out
All fuses out, circuit breakers open
Test:
   1 as shown for P–N
   2 dotted connections for PN–E
   Remove temporary link after PN–E test

Polarity tests

A major requirement for electrical installations is that single-pole switches and fuses should be connected in the phase conductor, and not in the neutral. Figure 14.19 shows how such a test should be carried out, a very low-resistance reading (usually a small fraction of an ohm) indicating correct connections. Care must be taken to ensure that the temporary link applied between the phase and protective systems is removed after testing.

Figure 14.19   Polarity testing

Earth tests

Tests for earth-fault-loop impedance and for earth electrodes were considered in Section 12.5.

14.8   Exercises

1 Write down a list of the items of equipment normally found at the incoming-supply position of an electrical installation, and state the purpose of each.

2 Explain why the main fuse and meter of an installation have seals applied so that interference with them can be detected.

3 Sketch a consumer's service unit, indicating its principal components.

4 Draw circuit diagrams of the wiring connections for
   (*a*) two lights controlled by one switch
   (*b*) one light controlled by either of two switches
   (*c*) two lights controlled by any one of three switches

5 A light in a bedroom is controlled by a one-way switch at the door. Using a wiring diagram, show how the system could be converted for control from the door or from a ceiling switch over the bed.

6 How many lighting points may be connected to one lighting circuit in compliance with IEE Wiring Regulations (*BS 7671*)? Would this number of lights normally be wired on one circuit? Give reasons for your answer.

7 Show the connections for two lights, each controlled by one switch, and two lights both controlled by two-way switches, if they are wired
   (*a*) using the looping-in method
   (*b*) in sheathed cable using joint boxes
   (*c*) in sheathed cable using three-plate ceiling roses.

8 Draw a circuit diagram to show the mains layout for a socket-outlet system comprising six 15 A and eight 5 A sockets.

9 What are the advantages of using socket outlets for which *only* fused plugs are available?

10 Draw a circuit showing the most economical method of wiring five 13 A sockets in a lounge of area 24 m^2.

11 (*a*) How many 13 A sockets may be fed from a ring circuit serving an area of 98 m^2?
   (*b*) How many spurs are allowed from such a ring circuit?
   (*c*) How many sockets may be connected to each spur?
   (*d*) How many fixed appliances may be connected to each spur?

12 What is a protective conductor? What forms may such a conductor take?

13 Why must single-pole switches be connected only in the phase conductor of a system having an earthed neutral? Describe a simple test to ensure that they are so connected.

14 Why must the continuity of a ring circuit be tested, and how can such a test be carried out?

15 Describe how a continuity test of a protective conductor would be carried out, and indicate the results which should be obtained.

16 What is insulation resistance? Describe an insulation-resistance test on a lighting circuit, giving the minimum acceptable result.

14.9 Multiple-choice exercises

14M1 The supply company's protection at the mains position is called the
   (*a*) circuit fuse   (*b*) service fuse
   (*c*) sealing box   (*d*) digital meter

14M2   A consumer's service unit does not usually contain a main fuse because
    (*a*) there is no possibility of a fault on the supply tails
    (*b*) leaving it out will save money
    (*c*) the service fuse will offer supply protection
    (*d*) there is seldom room for it within the unit

14M3   A split consumer's unit is likely to be used where
    (*a*) a larger unit is needed
    (*b*) ring circuits need to be separated from all others
    (*c*) all the circuit fuses required will not fit into a standard unit
    (*d*) some, but not all, of the circuits are to be RCD protected

14M4   When one or more RCDs are in use to provide local protection, the RCD installed at the mains position must be
    (*a*) time-delayed in operation
    (*b*) of larger current-carrying capacity
    (*c*) quicker in operation than the other RCDs
    (*d*) arranged to have exactly the same operating time as the other RCDs

14M5   It is good practice for at least two lighting circuits to be installed in every building so that
    (*a*) the installation will be more expensive
    (*b*) some of the lighting will still be available if one circuit is lost
    (*c*) more ways in the fuse board are used
    (*d*) all the lights will fail simultaneously

14M6   A special method of wiring two-way switching is advisable in situations where
    (*a*) the increased cost is acceptable
    (*b*) there is an increased likelihood of fire
    (*c*) lights must be controlled from more than one position
    (*d*) an induction-loop system is installed for people who are deaf

14M7   Three 5 A round-pin sockets may be wired on a single radial circuit when
    (*a*) the cable and the circuit protection are both rated at 15 A
    (*b*) the installation is carried out in a property belonging to the National Trust
    (*c*) they feed a floor area which does not exceed 20 m^2
    (*d*) they are to be used only with fused plugs

14M8   The floor area served by a radial circuit protected by a 32 A fuse must
    (*a*) not exceed 100 m^2 (*b*) not exceed 50 m^2
    (*c*) is unlimited   (*d*) not exceed 20 m^2

14M9   An unfused spur from a ring circuit for *BS 1363* sockets must
    (*a*) be wired in 4 mm^2 cable
    (*b*) be no longer than 18 m
    (*c*) feed as many sockets as required
    (*d*) feed no more than one twin or two single sockets

14M10   The number of sockets fed by unfused spurs from a ring circuit
    (*a*) is unlimited
    (*b*) must not be more than 10
    (*c*) must not exceed the number of sockets connected directly to the ring
    (*d*) must not be more than half those connected directly to the ring

14M11 The feed for a 3 kW immersion heater should be taken from
    (*a*) a lighting circuit
    (*b*) a separately 15 A protected circuit
    (*c*) a ring circuit as a spur
    (*d*) a radial 30 A circuit

14M12 One problem with an open-circuit fire- or burglar-alarm circuit is
    (*a*) that operating one contact alone will not cause the alarm to sound
    (*b*) the alarm sounder may fail prematurely
    (*c*) there is only one circuit for operation
    (*d*) failure of a circuit will not be apparent until the alarm is required

14M13 It is very important that, in an installation with an earthed neutral, single-pole switches and fuses
    (*a*) should be connected only in the phase conductor
    (*b*) should not be used at all
    (*c*) should be connected only in the neutral conductor
    (*d*) should be connected in both phase and neutral conductors

14M14 When protective-conductor continuity is confirmed with a resistance-measuring meter connected from phase to earth at each socket and a temporary link at the supply position from phase to earth
    (*a*) fuses must be in, circuit breakers and main switch closed
    (*b*) it makes no difference if fuses are in or out
    (*c*) all fuses must be out, circuit breakers off and main switch off
    (*d*) fuses must be out, but circuit breakers and main switch on

14M15 When a measurement of the resistance of a protective conductor is made with the aid of a wander lead, the resistance of this lead
    (*a*) is the resistance required
    (*b*) must be added to the measured value
    (*c*) should be ignored
    (*d*) must be subtracted from the measured value

14M16 The test voltage to be used for insulation resistance tests on a 240 V system will be
    (*a*) 500 V direct   (*b*) 1000 V direct
    (*c*) 500 V alternating   (*d*) 250 V direct

14M17 The minimum insulation resistance allowable by the IEE Wiring Regulations (*BS 7671*) is $0\cdot5$ MΩ. If the measured value for a small installation is $1\cdot0$ MΩ
    (*a*) the installation is entirely satisfactory
    (*b*) further investigation should be made to find the reason for the reading
    (*c*) the installation should be split into two and retested
    (*d*) the installation has failed the test

Chapter 15

Introduction to electronics

15.1   Introduction

Electronics is that branch of electrical engineering in which semiconductor devices are used. Not many years ago electronics was a separate subject, studied mainly by those interested in telecommunications, radio, television etc.

Today, electronics has reached into every field of electrical-engineering activity. Motor-control gear, thermostats, boiler-control systems, even the simple door bell, may, and probably do, contain electronic components. The electrical craftsman can no longer afford to be ignorant of electronics. This chapter will introduce the subject, and will describe some of the components used.

15.2   Resistors for electronic circuits

Most of the resistors considered in Chapter 2 are types used in power circuits, and must be capable of carrying heavy currents. In electronic circuits, we are concerned usually with currents of the order of milliamperes, so we are able to accept resistors with low current ratings. Since the currents are low, and because large numbers of resistors are needed, it becomes economical to use small resistors whose value is not so accurately known as is the case with power resistors. For example, a 120 Ω marked 10% resistor may have a value anywhere between 108 Ω and 132 Ω. This variation in values is called the **tolerance**, and is usually given as a percentage (here ± 10%).

Example 15.1

What are the maximum and minimum acceptable values for a resistor marked at 15 kΩ if its tolerance is

  (a) ±20%,   (b) ±10%,   (c) ±5%,   (d) ±2%,   e) ±1%?

(a) 20% of 15 000 Ω is $\dfrac{20}{100} \times 15\ 000$ ohms = 3000 Ω or 3 kΩ

Maximum value is thus 15 kΩ + 3 kΩ = 18 kΩ
Minimum value is thus 15 kΩ – 3 kΩ = 12 kΩ

(b) 10% of 15 000 Ω is $\dfrac{10}{100} \times 15\ 000$ ohms = 1500 Ω or 1·5 kΩ

Maximum value is thus 15 kΩ + 1·5 kΩ = 16·5 kΩ
Minimum value is thus 15 kΩ – 1·5 kΩ = 13·5 kΩ

(c) 5% of 15 000 Ω is $\dfrac{5}{100} \times 15\ 000$ ohms = 750 Ω or 0·75 kΩ

Maximum value is thus 15 kΩ + 0·75 kΩ = 15·75 kΩ
Minimum value is thus 15 kΩ – 0·75 kΩ = 14·25 kΩ

(d) 2% of 15 000 Ω is $\dfrac{2}{100} \times 15\ 000$ ohms = 300 Ω or 0·3 kΩ

Maximum value is thus 15 kΩ + 0·3 kΩ = 15·3 kΩ
Minimum value is thus 15 kΩ – 0·3 kΩ = 14·7 kΩ

(e) 1% of 15 000 Ω is $\dfrac{1}{100} \times 15\ 000$ ohms = 150 Ω or 0·15 kΩ

Maximum value is thus 15 kΩ + 0·15 kΩ = 15·15 kΩ
Minimum value is thus 15 kΩ – 0·15 kΩ = 14·85 kΩ.

A very important property of a resistor is its **power rating**. This is the power which may be dissipated in the resistor continuously without it becoming overheated, and depends on the current carried:

$$P = I^2 R \ \text{ so } \ I^2 = \frac{P}{R}$$

and
$$I = \sqrt{\frac{P}{R}}$$

where I = maximum sustained permissible current, A
$\qquad P$ = power rating of resistor, W
$\qquad R$ = resistance of resistor, Ω.

Example 15.2

Calculate the maximum permissible current in a 1 kΩ resistor if it is rated at (a) 0·5 W (b) 1 W (c) 2 W.

(a)
$$I = \sqrt{\frac{P}{R}} = \sqrt{\frac{0\cdot5}{1000}} = \sqrt{0\cdot0005} = 0\cdot0224 \text{ A}$$
$$\text{or } 22\cdot4 \text{ mA}$$

(b)
$$I = \sqrt{\frac{P}{R}} = \sqrt{\frac{1}{1000}} = \sqrt{0\cdot001} = 0\cdot0316 \text{ A}$$
$$\text{or } 31\cdot6 \text{ mA}$$

(c)
$$I = \sqrt{\frac{P}{R}} = \sqrt{\frac{2}{1000}} = \sqrt{0\cdot002} = 0\cdot0447 \text{ A}$$
$$\text{or } 44\cdot7 \text{ mA}$$

Example 15.3

What rating should be chosen for a 12 kΩ resistor which is to carry a current of 9 mA?

$$P = I^2R = \left(\frac{9}{10^3}\right)^2 \times 12 \times 10^3 \text{ watts}$$

$$= \frac{81}{10^6} \times 12 \times 10^3 \text{ watts}$$

$$= \frac{972}{10^3} \text{ watts}$$

$$= 0\cdot972 \text{ W}$$

In practice, the nearest rating above the calculated value would be chosen, i.e. 1 W.

The resistor is often too small for its value and power rating to be printed on it, so a code is used to show its resistance. Two codes are commonly in use:

(a) *Colour code:* A series of bands of colour are printed on the resistor, each colour representing a number, or in some cases a tolerance. The colours, with the values represented, are

black	0
brown	1
red	2
orange	3
yellow	4
green	5
blue	6
violet	7
grey	8
white	9

The colours are applied in four bands (Figure 15.1). The first band indicates the first figure of the value, the second band the second figure, and the third band the number of noughts to be added. The third band is sometimes coloured gold or silver, indicating one-tenth or one-hundredth, respectively, of the first two digits. Thus when bands appear in the sequence blue-grey-red (in positions 1, 2 and 3, respectively, in Figure 15.1), the resistor is 6800 Ω or 6·8 kΩ. Similarly, orange-white-green indicates 3 900 000 Ω, or 3·9 MΩ, and red-red-gold 2·2 Ω.

bands

Figure 15.1   Colour coding of resistors

The fourth band indicates the tolerance, the colour code being

gold ± 5% ⎞   not to be confused with gold (× 1/10) and
silver ± 10% ⎠   silver (× 1/100) when used as the third band.

No fourth band or a salmon-pink band indicates ± 20%.

If the bands 1, 2, 3 and 4 are, respectively, red-red-brown-silver, it indicates 220 Ω ± 10%. Sometimes a resistor is specially made to ensure that its resistance does not change as it ages. These 'high-stability' resistors are indicated by a salmon-pink fifth band.

(*b*) *Digital code:* This code takes the form of numbers and letters printed on the resistor, and is most often applied to wire-wound types.

R indicates a decimal point, so that 1R0 means 1Ω, 4R7 means 4·7 Ω, 68 R means 68Ω, 220R means 220 Ω, and so on.

K has a similar function, but indicates values in thousands of ohms, or kilohms, so that 1K0 means 1 kΩ, 4K7 means 4·7 kΩ, 82K means 82 kΩ, and so on.

M serves a similar purpose, but indicates values in millions of ohms, or megohms. Thus 1M2 means 1·2 MΩ, 15M means 15 MΩ, and so on.

With this code, tolerances are indicated by a code letter placed after the value.

The tolerance code is

B ± 0·1%
C ± 0·25%
D ± 0·5%
F ± 1%
G ± 2%
J ± 5%
K ± 10%
M ± 20%
N ± 30%

Examples of the use of this code are

4R7J   = 4·7 Ω ± 5%
6K8F   = 6·8 kΩ ± 1%
68KK   = 68 kΩ ± 10%
4M7M = 4·7 MΩ ± 20%

Many types of resistor are used in electronic circuits:

Carbon-composition resistors are most usually used, and are moulded from carbon compound into a cylindrical shape, connection being by wire ends. They are made in various sizes with power ratings from ⅛ W to 2 W (Figure 15.2).

Carbon-film resistors often have higher stability than moulded-carbon types. The resistive film or coating is deposited on a glass tube, which is buried in a plastics moulding. The connecting wires are carried into the ends of the glass tube to conduct heat away from the resistor.

Cracked-carbon film, or pyrolytic resistors have a film of cracked carbon deposited on a ceramic rod. The film is then cut through in a spiral pattern, producing what is effectively a long thin resistor element wound round the rod. Endcaps with connecting leads, and a coating of silicon lacquer, complete the construction. Cracked-carbon resistors have higher stability than other types.

Metal-film and metal-oxide resistors have a similar construction to the cracked-carbon film type, but the film, and hence the spiral track, is formed of nickel–chromium or a metal oxide. These resistors are capable of operation at very high temperatures.

Wire-wound resistors are used when the power to be dissipated is high. The resistance wire (nickel–chromium) is usually wound on a ceramic tube and given a vitreous-enamel coating for insulation and protection (Figure 15.3).

Variable resistors are usually operated by turning a shaft, and are of two types. The wire-wound type (see Figure 2.1*c* for a much higher-power type) is used for lower resistances (up to about 100 kΩ) where higher powers

Figure 15.2 Carbon-composition resistors
a Uninsulated-type
b Insulated-type

Figure 15.3 Wire-wound vitreous-enamelled resistor

are dissipated. The carbon-track type is used for higher resistances (up to about 2 MΩ).

15.3 Semiconductor diode

A crystal of the basic semiconductor material, germanium or silicon, has opposite sides treated so that the two halves have different characteristics. One-half of the crystal becomes a *p*-type material, with a shortage of mobile electrons, and the other half becomes an *n*-type material, which has a surplus of mobile electrons. Some of these surplus electrons cross the boundary from the *n*-type half to the *p*-type half of the crystal, making the *p*-type material negatively charged, and the *n*-type material positively charged (Figure 15.4).

Electrons can move more easily from the *n*-type region to the *p*-type, but not in the reverse direction. The device thus behaves as a rectifier. The circuit symbol for the semiconductor diode is shown in Figure 15.5, the arrow of the symbol pointing the direction in which **conventional** current can flow.

Figure 15.6 shows a typical rectifying circuit, and indicates how current can pass through the load during the positive halfcycles of an alternating supply, but not during the negative halfcycles.

In the direction in which current flows easily (the **forward direction**) there will be no current through the diode until a certain applied voltage

Figure 15.4 Simple representation of *p–n* junction (semiconductor diode)

Figure 15.5   Circuit symbol for semiconductor diode

Figure 15.6   Circuit and wave diagrams for semiconductor diode connected to load as halfwave rectifier

is reached. This is the **forward volt drop** which is about $0\cdot3$ V for germanium diodes and about $0\cdot7$ V for silicon types. In the **reverse direction** the diode will not allow current to flow at all (other than a very small leakage current) unless the applied voltage is large enough to break down the *p–n* junction. When this happens the diode is destroyed as a rectifier; the **reverse-breakdown voltage** varies with the construction of the diode, but is seldom less than 25 V for germanium and 75 V for silicon types.

Semiconductor diodes are used widely in industry, their main application being the conversion of an AC supply to give a DC supply. Although the action of a single diode does give a direct current, this consists of a series of isolated current pulses, and is quite unlike the output from a DC generator or a battery. By using two or more diodes in special circuits, the output may be improved, and the addition of capacitors and chokes allows a smooth supply to be obtained.

15.4 Semiconductor-diode types

There are two main types of semiconductor diode, each having its particular advantages and disadvantages.

Germanium diodes have a high current rating for a comparatively small size, but are less widely used, as failure occurs suddenly if overload raises the junction temperature above about 80°C.

Silicon diodes have a very high current rating indeed, and an extremely high reverse breakdown voltage. A silicon diode no larger than a pea can carry a current of 15 A, and one 50 mm in diameter and 40 mm long can handle over 600 A. Figure 15.7 shows a typical construction for a silicon diode.

Figure 15.7   Silicon-diode construction

These rectifiers must be kept cool to prevent a breakdown, and are often mounted on aluminium castings called **heatsinks** to increase current rating. Very heavily loaded silicon diodes are sometimes water-cooled.

15.5 Transistors

The diode was the first electronic component we encountered, and it is known as a **passive device**, because under normal circumstances it is not possible to change its effective operation. The transistor, however, is an

active device because its operation can be controlled by the application of a current or a voltage.

There are very many types of transistors, but we will limit our explanation to just two of the most common types.

Bipolar junction transistor (BJT)

This device is still probably the most common transistor, especially when used in amplifiers. It is composed of three layers of semiconducting material, either a very thin layer of *p*-type sandwiched between two thicker layers of *n*-type (the *npn* transistor), or two thick layers of *p*-type separated by a thin layer of *n*-type (the *pnp* transistor), although this latter type is now uncommon. An explanation of *p*-type and *n*-type semiconductor materials was given in Section 15.3. In both cases, the thin central layer is called the **base**, whilst the two thicker outer layers are known as the **emitter** and the **collector**. The physical arrangements of the two types, together with their circuit symbols, are shown in Figure 15.8.

The operation of the BJT transistor is complicated, but a simplified explanation of how the *npn* type works is as follows. First we must remind ourselves of the operation of the diode, because the BJT transistor is in some respects like two diodes connected together. With the diode, we can consider that in the forward direction, in which current flows more easily, the device will have low resistance. When the applied voltage is such that the diode behaves in this way, it is said to be **forward biased** and in this condition its resistance is comparatively low. If the applied voltage charge is reversed the diode is said to be **reverse biased**, in which condition it has very high resistance and hardly any current flows.

The BJT transistor has two junctions, that between the emitter and base (the e–b junction) and that between the collector and base (the c–b junction). The transistor is connected in circuit as shown in Figure 15.9, the applied voltages causing the e–b junction to be forward biased so that its resistance is low and the c–b junction reverse biased, resulting in very high resistance.

The amplifying action of the transistor may be simply explained by reference to Figure 15.10. Current (in the form of a flow of electrons) enters the base region from the emitter, driven by the emitter–base voltage against the low resistance of the e–b junction. Because the resistance is low, a small change in voltage will result in a large current change.

As the base region is very thin, most of the electrons entering from the emitter are unable to leave by the base connection, but 'blunder' across the high-resistance collector–base (c–b) junction. Thus, a small voltage change in the emitter–base low-resistance circuit results in a significant current change in the collector–base high-resistance circuit.

As the *c–b* junction has such high resistance, the current flow is hardly affected by the connection of a high-value resistor (usually 10–20 kΩ) in the collector circuit (the collector load). Thus, a small voltage change in the emitter–base circuit results in a significant current change in the collector

(i)

(ii)

Figure 15.8 *npn* **and** *pnp* **types of bipolar junction transistors**
(i) *npn*-transistor arrangement and circuit symbol
(ii) *pnp*-transistor arrangement and circuit symbol

Figure 15.9 **Biasing voltages for the** *npn* **bipolar junction transistor**
Emitter-base (e–b) junction forward biased – low resistance
Collector-base (c–b) junction reverse biased – high resistance

Figure 15.10   Explanation of amplification operation

load resistor and hence a much higher change in the voltage across it. Hence the circuit has performed as a voltage amplifier.

It must be appreciated that this description is grossly oversimplified, but hopefully will give some inkling of how the device performs its function. The BJT can be considered as a resistor whose resistance is reduced as the base current increases and increases as the base current becomes smaller. It is called an active device because the circuit resistance is controlled by electrical means rather than by physical movement, such as the operation of a slider.

Figure 15.11   Transistor as a voltage amplifer

Field-effect transistor (FET)

This type of transistor is used very widely indeed in data and logic circuits, such as those used in computers, in programmable controllers and so on. Like the BJT, the FET is an active device whose resistance is controlled by the application of a variable low voltage to the gate connection. The two most important types of FET are the **junction-gate type (JUGFET)** and the **insulated-gate type (IGFET)**.

JUGFET

This transistor can be made as *n*-channel or *p*-channel type, the former being more common. Circuit symbols for the two types are shown in Figure 15.12.

The *n*-channel type consists of a cylinder of *n*-type silicon surrounded by a *p*-type shroud or tube, so that the conducting channel between the end contacts is through the *n*-type material. The general arrangement is shown in Figure 15.13, which also makes it clear that the connections to the ends of the channel are called the **source** (S) and the **drain** (D), while the connection to the surrounding cylinder is called the **gate** (G).

Figure 15.12   Circuit symbols for JUGFETS
(i) *n*-channel type
(ii) *p*-channel type

Figure 15.13   Arrangement of an *n*-channel JUGFET

If the gate is made negative, a depletion layer is formed within the silicon tube as shown in Figure 15.14. This reduces the cross-sectional area of the channel available for electron flow, and thus reduces the current.

If the negative potential on the gate is increased further, a point will be reached where the depletion layers meet, completely closing the *n*-channel. This is called the **pinch-off** point and no current flows in the device. This condition is illustrated in Figure 15.15.

The simplified account above makes it clear that the current flowing in the field-effect transistor can be controlled by the level of the negative voltage applied to the gate. In most cases, FETs are used at voltages not exceeding 5 V. The *p*-channel-type JUGFET operates in much the same way as the *n*-channel-type, except that the tube is *n*-type and the cylinder is *p*-type.

Figure 15.14   Negative gate potential reduces the cross-sectional area of the current path

Figure 15.15   Increased negative gate voltage has widened the depletion layers, 'pinching off' the channel

IGFET

As the name 'insulated-gate field-effect transistor' indicates, this device has its gate connection insulated from the channel by a layer of silicon dioxide. It again exists in *n*-channel and *p*-channel forms, circuit symbols being shown in Figure 15.16 and a simplified construction in Figure 15.17.

As in the JUGFET, pinch-off can be achieved by making the insulated gate negative with respect to the source, when the transistor is said to be **depleted**. However, in this type of transistor, the opposite effect is also possible, a positive voltage on the gate increasing the current in the channel, when the transistor is said to be **enhanced**. The connection to the substrate is available and may be used as the other terminal for biasing the gate, although the substrate is often connected to the source. Because of the metal oxide and silicon used in construction, the IGFET is sometimes referred to as a **MOSFET**.

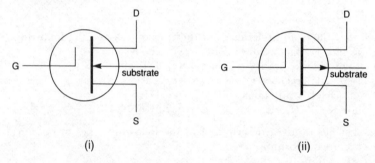

Figure 15.16   circuit symbols for IGFETS
    (i) *n*-channel type
    (ii) *p*-channel type

Figure 15.17   Simplified construction of an IGFET

15.6   Exercises

1 What are the possible maximum and minimum values of a resistor marked
680 Ω if its tolerance is
(a) ±20%
(b) ±10%
(c) ±5%
(d) ±2%?

2 A resistor is marked as 180 kΩ but measured at 192 kΩ. What tolerance does
this indicate?

3 Calculate the maximum currents for the following resistors:
(a) 2·7 kΩ, 0·5 W
(b) 68 Ω, 0·125 W
(c) 120 kΩ, 1 W
(d) 1·2 MΩ, 0·5 W
(e) 43 kΩ, 2 W.

4 Give the values and tolerances of the resistors having the following codes:
(a) orange, orange, yellow, gold   (f) 2K2K
(b) brown, black, orange, silver   (g) 3M9M
(c) grey, red, green   (h) 39RG
(d) red, violet, gold, silver   (i) 47KJ
(e) 390RK   (j) 5K6N

5 Use sketches to assist descriptions of the following types of resistor:
(a) carbon-composition
(b) carbon-film
(c) pyrolytic
(d) metal-oxide
(e) wire-wound.

6 Draw a circuit diagram to show a semiconductor diode connected in series
with a load to an AC supply. Draw a wave diagram of the supply voltage and
the circuit current, describing why the current wave has the form shown.

7 Sketch the circuit symbol of a *pnp* bipolar-junction transistor, labelling the
connections.
    With the aid of a sketch showing its construction, explain why a small voltage
change across the low-resistance base–emitter region of an *npn* BJT will result
in a much larger than expected current change in the high-resistance
collector–base region.

8 Draw a circuit diagram of an *npn* bipolar-junction transistor connected for
use as a voltage amplifier.

9 Explain in simple terms the construction and the operation of a junction-gate
field-effect transistor (JUGFET). In the course of your explanation, indicate
the meaning of the terms **depletion layer** and **pinch-off**.

10 Draw the circuit symbols for both *n*-channel and *p*-channel insulated-gate
field-effect transistors (IGFETs), labelling the connections.

15.7 Multiple-choice exercises

15M1 A resistor marked as having a value of 47 kΩ ± 10% may have an actual value between
(*a*) 42·3 kΩ and 47 kΩ   (*b*) 42·3 kΩ and 51·7 kΩ
(*c*) 37 kΩ and 57 kΩ   (*d*) 44·7 kΩ and 49·4 kΩ

15M2 A suitable power rating for a 68 kΩ resistor carrying a current of 3·5 mA is
(*a*) 250 W   (*b*) 0·1 W   (*c*) 2 W   (*d*) 1 W

15M3 A 1·2 MΩ resistor with a power rating of 0·5 W should carry no more current than
(*a*) 0·65 mA   (*b*) 0·42 μA   (*c*) 0·65 A   (*d*) zero

15M4 The coloured band printed on a resistor to represent the number 6 is
(*a*) red   (*b*) violet   (*c*) yellow   (*d*) blue

15M5 The third coloured band on a resistor represents
(*a*) the power rating of the resistor
(*b*) the tolerance of the resistor
(*c*) the number of zeros to be added to the value given by the first two colour bands
(*d*) the temperature at which the resistor will safely operate

15M6 If the four coloured bands on a resistor are, from left to right, brown, grey, yellow and silver, the resistor value and tolerance will be
(*a*) 180 kΩ ± 10%   (*b*) 18 MΩ ± 10%
(*c*) 27 kΩ ± 5%   (*d*) 18 kΩ ± 10%

15M7 A 56 Ω resistor with a tolerance of ± 5% is indicated by bands with colours
(*a*) orange–blue–grey–silver
(*b*) green–blue–black–gold
(*c*) blue–green–black–silver
(*d*) green–blue–black–salmon pink

15M8 If a resistor is marked 1M8M, its value and tolerance are
(*a*) 18 MΩ ± 10%   (*b*) 1·8 MΩ ± 5%
(*c*) 1·8 kΩ ± 20%   (*d*) 1·8 MΩ ± 20%

15M9 The digital marking of 680KJ on a resistor indicates a value and tolerance of
(*a*) 680 kΩ ± 5%   (*b*) 68 kΩ ± 10%
(*c*) 6·8 kΩ ± 10%   (*d*) 68 kΩ ± 20%

15M10 A resistor of 33 kΩ with a tolerance of ± 10% will carry the digital marking
(*a*) 33KJ   (*b*) 3K3K   (*c*) 33MK   (*d*) 33KK

15M11 The type of resistor to select if the power it is to dissipate will be 15 W will be
(*a*) wirewound   (*b*) cracked carbon film
(*c*) carbon composition   (*d*) metal-oxide

15M12 The arrow in the symbol used for a semiconductor diode points in the
direction of
(*a*) easy electron movement
(*b*) the positive connection
(*c*) easy conventional current flow
(*d*) the connection to the supply

15M13 A semiconductor material with mobile electrons and an overall positive
charge is called
(*a*) a *p*-type material   (*b*) a rectifier
(*c*) an *n*-type material   (*d*) a diode

15M14 The diagram below is for a
(*a*) power resistor   (*b*) halfwave rectifier
(*c*) heating circuit   (*d*) full-wave rectifier

15M15 The most common material used in semiconductor devices is
(*a*) copper oxide   (*b*) germanium
(*c*) selenium   (*d*) silicon

15M16 The three regions of the bipolar-junction transistor are called the
(*a*) emitter, base and collector
(*b*) diode, collector and base
(*c*) emitter, forward bias and collector
(*d*) source, gate and drain

15M17 The emitter–base (*e–b*) junction of a bipolar-junction transistor is
(*a*) reverse biased to have very high resistance
(*b*) forward biased to have low resistance
(*c*) extremely thin
(*d*) composed of a *p*-type silicon cylinder

15M18 The circuit symbol for an *npn* bipolar transistor is

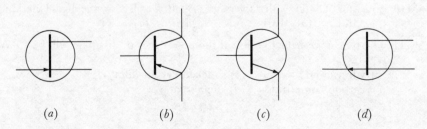

(*a*)   (*b*)   (*c*)   (*d*)

15M19 The three connections of a junction-gate field-effect transistor are called the
    (*a*) emitter, base and collector
    (*b*) source, depletion layer and gate
    (*c*) source, gate and drain
    (*d*) positive and negative

15M20 When the depletion layers in the channel of a field-effect transistor meet, the device is said to be
    (*a*) nonconducting   (*b*) switched off
    (*c*) closed   (*d*) pinched-off

15M21 The circuit symbol shown represents

    (*a*) a silicon diode
    (*b*) a *p*-channel-type insulated-gate field-effect transistor
    (*c*) an *npn* bipolar-junction transistor
    (*d*) an *n*-channel-type junction-gate field-effect transistor

Numcrical answers to exercises

Chapter 1

Exercises (Section 1.14)

1 1200 C

2 100 min

3 12·5 A

5 900 J

6 25 mm

7 150 N

8 44 J

9 120 000 J

10 2·89 μV

11 0·5 A

12 60 A

13 4 mA

14 20 Ω

15 10 V

16 15 A

17 3 A, 15 V, 45 V, 60 V, 120 V

18 2 Ω

19 479·8 Ω

20 (a) 1·5 Ω,

(b) 3·6 MΩ

(c) 80 $\mu\Omega$

21 (a) overcurrent, heating

(b) coulomb

(c) copper, aluminium

(d) the bell, the motor

(e) PVE, mica

(f) negative, positive

(g) heating

(h) negative

(i) 6 Ω

(j) 5 Ω

(k) current × time

(l) 360 mA

(m) 3 300 V

22 30 V, 3 A, 3 Ω

23 0·067 Ω, 6 000 A 4 000 A, 1 333 A, 667 A

24 12 V

25 (a) see Figure 1.11

(b) one-half, one-quarter

26 10 A, 6 A, 4 A, 20 A, 2·4 Ω

27 (a) 78 V (b) 312 V

28 (b) 4·8 A (c) 41·67 Ω

29 (a) 80 Ω

(b) $I_1 = I_2 = I_3 = 1$ A

$I_4 = 2\cdot25$ A

$I_5 = 0\cdot75$ A

$I_6 = 1\cdot8$ A

$I_7 = 0\cdot9$ A

$I_8 = 0\cdot3$ A

(c) $U_1 = U_2 = U_3 = 60$ V

$U_4 = U_5 = 90$ V

$U_6 = U_7 = U_8 = 90$ V

30 6 Ω

31 7·8 Ω

32 Bank 1: 30 V, 0·75 A, 1·25 A

Bank 2: 10 V, 2 A

Bank 3: 8 V, 0·5 A in each 16 Ω resistor

33 10 A

Multiple-choice answers (Section 1.15)

1M1 (b)

1M2 (a)

1M3 (d)

1M4 (d)

1M5 (b)

1M6 (d)

1M7 (a)

1M8 (c)

1M9 (a)

1M10 (d)

1M11 (b)

1M12 (a)

1M13 (b)

1M14 (d)

1M15 (b)

1M16 (c)

Chapter 2
Exercises (Section 2.9)

1 $2 \cdot 16$ Ω
2 $2 \cdot 5$ Ω
3 $0 \cdot 194$ Ω
4 $0 \cdot 1187$ Ω
5 $2 \cdot 15$ Ω
6 250 m
7 2mm^2
8 $0 \cdot 75$ Ω
9 $2 \cdot 5$ times
10 100 Ω
13 $0 \cdot 342$ Ω

14 $3 \cdot 82$ Ω
15 $1 \cdot 72$ Ω
16 116 m
17 2500 μΩmm
18 $13 \cdot 3$ mm^2
19 439 m
20 $\rho = 16 \cdot 8$ μΩm, so the cable is probably of copper
21 $6 \cdot 57$ mm^2
22 $109 \cdot 9$ Ω

23 $1 \cdot 64$ Ω
24 $15 \cdot 06$ Ω
25 188°C
26 200°C
27 $3 \cdot 7$ Ω
28 $12 \cdot 8$ Ω
29 $17 \cdot 0$ Ω
30 $9 \cdot 8$ V
31 $2 \cdot 15$ A
32 $3 \cdot 56$ Ω, $4 \cdot 41$ Ω
33 $0 \cdot 1\%$

34 (a) 192 V (b) 221 V (c) 230 V (d) 240 V
35 3 V
36 236 V
37 $0 \cdot 40$ Ω
38 $4 \cdot 31$ A
39 234 V
40 $16 \cdot 8$ mm^2
41 35 m
42 $195 \cdot 9$ V

Multiple-choice answers (Section 2.10)

2M1 (b)   2M2 (c)   2M3 (a)   2M4 (d)   2M5 (a)
2M6 (c)   2M7 (b)   2M8 (d)   2M9 (a)   2M10 (b)
2M11 (d)   2M12 (c)

Chapter 3
Exercises (Section 3.7)

1 19 620 N
2 $51 \cdot 0$ kg
3 200 N
4 (a) 60 000 N/m^2 (b) 800 N/m^2
5 $66 \cdot 7$ N

6 20 Nm
7 4200 J, 35 W
8 22 500 J
9 4 kW
10 480 N, $2 \cdot 4$ N/mm^2
11 600 N, 9

12 375 N
13 160 N
14 $0 \cdot 139$ m, 222 Nm
15 500 mm, 500 Nm
16 130 teeth, 446 r/min

17 2800 N, 82°
18 127°from 6000 N, 10 000 N
19 600 N
20 (b) 1156 N (c) 1250 N

Multiple-choice answers (Section 3.8)

3M1 (c)   3M2 (a)   3M3 (d)   3M4 (a)   3M5 (b)
3M6 (d)   3M7 (c)   3M8 (b)   3M9 (a)   3M10 (d)
3M11 (d)   3M12 (b)   3M13 (a)   3M14 (c)   3M15 (b)
3M16 (c)

Chapter 4
Exercises (Section 4.8)

1 (a) 333 K (b) 198 K (c) 1273 K
2 (a) 47°C (b) 1227°C (c) $- 33$°C

3 $2 \cdot 68$ MJ
4 $3 \cdot 76$ MJ, $1 \cdot 04$ kWh
5 $9 \cdot 01$ kWh
6 $64 \cdot 1$ MJ
7 $16 \cdot 7$ MJ

8 8370 J/s
9 1100 kWh
10 $34 \cdot 9$°C
11 120°C
12 $14 \cdot 5$°C

13 27 min 55 s
14 $10 \cdot 5$ kW
15 369°C

Multiple-choice answers (Section 4.9)

4M1 (c)   4M2 (a)   4M3 (d)   4M4 (b)   4M5 (b)
4M6 (a)   4M7 (d)   4M8 (b)   4M9 (c)

Chapter 5

Exercises (Section 5.5)

1 1·5 kW	12 1·33 kW	21 4·5 kW, £10·80	29 2 A,
2 48 W	13 200 kWh or	22 5·5 Ω	10 A, 30 A
3 500 W	720 MJ	23 $P_7 = 175$ W	27 £9·46
4 2·4 kW	14 15 000 J	$P_5 = 45$ W	28 1 kW
5 40 Ω	15 12 V	$P_{10} = 22·5$ W	29 54 min 16 s
6 10 kΩ	16 £4·80	$P_{30} = 7·5$ W	30 2·62 kW
7 3·125 A	17 744 W	24 20 Ω	31 16·4 A
8 10 A	18 2 kW	25 (a) 12 A, 6 A	32 80%
9 300 V	19 2·7 kW	(b) 4·32 kW	33 21·6 kW
10 7·07 mA, 70·7 V	20 20 kW, 83·3 A	26 6·33 A,	34 6 kW, 11·8 kW
11 300 J			

Multiple-choice answers (Section 5.6)

5M1 (c)	5M2 (a)	5M3 (d)	5M4 (a)	5M5 (b)
5M6 (c)	5M7 (b)	5M8 (d)	5M9 (a)	5M10 (b)
5M11 (c)	5M12 (b)	5M13 (d)	5M14 (c)	

Chapter 6

Exercises (Section 6.8)

4 1·3 T	(c) 6·03 mT	9 0·462 mWb	(c) decrease
5 6·72 μWb	7 2120 At	11 (a) increase	(d) decrease
6 (a) 4800 At/m	8 965 μWb, 1·21 T	(c) increase	(e) increase
(b) 4·82 μWb			

Multiple-choice answers (Section 6.9)

6M1 (b)	6M2 (d)	6M3 (a)	6M4 (b)	6M5 (c)
6M6 (a)	6M7 (d)	6M8 (b)	6M9 (b)	6M10 (d)
6M11 (c)	6M12 (a)	6M13 (b)	6M14 (d)	

Chapter 7

Multiple-choice answers (Section 7.10)

7M1 (c)	7M2 (b)	7M3 (d)	7M4 (a)	7M5 (c)
7M6 (b)	7M7 (b)	7M8 (a)	7M9 (d)	7M10 (c)
7M11 (a)	7M12 (c)			

Chapter 8

Exercises (Section 8.9)

4 1·4 V	11 3 A, 5·4 V	17 0·4 Ω	22 (a) (i) 1·12 A
5 2 A, 2·1 V	13 27 V	18 1·05 V	(ii) 8·06 V
6 0·2 Ω	14 (a) 2·3 V (b) 3·1 V	19 (a) 90 V	(b) (i) 0·193 A
7 0·0178 Ω	15 (b) (i) 5·5 V, 2 A	(b) 110 V	(ii) 1·39 V
8 0·025 Ω	(ii) 1·1 V, 10 A	20 1·4 A, 1·26 V	(c) (i) 0·549 A
10 12 V	16 36 V	21 0·4 Ω, 8 V	(ii) 3·95 V
			23 80%, 66·7%

Multiple-choice answers (Section 8.10)

8M1 *(c)*	8M2 *(b)*	8M3 *(a)*	8M4 *(d)*	8M5 *(b)*
8M6 *(d)*	8M7 *(a)*	8M8 *(d)*	8M9 *(c)*	8M10 *(c)*
8M11 *(a)*	8M12 *(b)*	8M13 *(d)*	8M14 *(a)*	

Chapter 9
Exercises (Section 9.8)

1 *(a)* into paper	*(d)* right to left	4 2 m/s	7 4 V
(b) out of paper	2 2 m	5 0·48 V	8 0·167 s
(c) north pole at bottom	3 3 m/s	6 4 V, 800 V	9 60 mWb

Multiple-choice answers (Section 9.9)

9M1 *(b)*	9M2 *(d)*	9M3 *(a)*	9M4 *(c)*	9M5 *(a)*
9M6 *(d)*	9M7 *(a)*	9M8 *(d)*	9M9 *(b)*	9M10 *(a)*
9M11 *(a)*	9M12 *(d)*	9M13 *(c)*		

Chapter 10
Exercises (Section 10.11)

1 1000 Hz	*(e)* – 40 A	*(c)* 50 A *(d)* 10 A	*(c)* 17·7 Ω, 0·17 A
2 16·7 ms	7 127 V, 141 V, 1·11	*(e)* 15 A	*(d)* 3·18 MΩ,
3 *(a)* 33·3 Hz	8 9·01 A, 14·1 A	17 *(a)* 314·2 Ω,	7·54 µA
(b) 150 V	9 216 V, 339 V	0·765 A	*(e)* 3·98 Ω, 0·251 A
(c) 162 V	10 191 A, 212 A	*(b)* 7·54 Ω,	20 100 turns
(d) 1·08	11 *(b)* 70·7A	55·4 A	21 125 V
4 141 V, 127 V	12 141 A	*(c)* 377 Ω,	22 1920 turns
5 20 ms	13 *(a)* 141 A, *(b)* 135°	0·265 A	25 *(a)* 480 turns
6 *(a)* 25 Hz	14 24·3 A leading by	18 *(a)* 318 Ω,	*(b)* 24 V
(b) 35 A	17°	0·754 A	27 4·17 A
(c) 38·7 A	15 100 V, 60°	*(b)* 7·96 kΩ,	28 100 A
(d) 1·105	16 *(a)* 12 A *(b)* 5·75 A	1·51 mA	29 2500 turns, 1·5 kV

Multiple-choice answers (Section 10.12)

10M1 *(b)*	10M2 *(d)*	10M3 *(a)*	10M4 *(c)*
10M5 *(a)*	10M6 *(d)*	10M7 *(a)*	10M8 *(b)*
10M9 *(c)*	10M10 *(b)*	10M11 *(c)*	10M12 *(b)*
10M13 *(d)*	10M14 *(a)*	10M15 *(d)*	10M16 *(c)*
10M17 *(b)*	10M18 *(a)*	10M19 *(b)*	10M20 *(d)*
10M21 *(c)*	10M22 *(b)*	10M23 *(a)*	10M24 *(c)*
10M25 *(b)*	10M26 *(a)*		

Chapter 11
Exercises (Section 11.8)

1 4·44 A	7 *(a)* right to left	8 1·11 T
2 13·3 m	*(b)* right to left	9 20 m
3 1·1T	*(c)* out of the paper	10 *(b)* 12 N
5 4000 n	*(d)* north pole at bottom	
6 0·25 mA		

Multiple-choice answers (Section 11.9)

11M1 *(c)*	11M2 *(d)*	11M3 *(a)*	11M4 *(d)*
11M5 *(b)*	11M6 *(c)*	11M7 *(a)*	11M8 *(c)*
11M9 *(b)*	11M10 *(d)*		

Chapter 12

Multiple-choice answers (Section 12.16)

12M1 *(c)*	12M2 *(d)*	12M3 *(a)*	12M4 *(c)*
12M5 *(b)*	12M6 *(a)*	12M7 *(d)*	12M8 *(b)*
12M9 *(d)*	12M10 *(a)*	12M11 *(c)*	12M12 *(b)*
12M13 *(a)*	12M14 *(c)*	12M15 *(d)*	12M16 *(a)*
12M17 *(a)*	12M18 *(d)*	12M19 *(b)*	12M20 *(c)*
12M21 *(d)*	12M22 *(a)*		

Chapter 13

Multiple-choice answers (Section 13.14)

13M1 *(b)*	13M2 *(d)*	13M3 *(a)*	13M4 *(d)*
13M5 *(c)*	13M6 *(a)*	12M7 *(b)*	13M8 *(d)*
13M9 *(a)*	13M10 *(c)*	13M11 *(c)*	13M12 *(b)*
13M13 *(a)*	13M14 *(d)*	13M15 *(a)*	

Chapter 14

Exercises (Section 14.8)

6 36, no

11 *(a)* unlimited

(b) not more than the number of sockets on the ring itself

(c) two

(d) one

Multiple-choice answers (Section 14.9)

14M1 *(b)*	14M2 *(c)*	14M3 *(d)*	14M4 *(a)*
14M5 *(b)*	14M6 *(d)*	14M7 *(a)*	14M8 *(b)*
14M9 *(d)*	14M10 *(c)*	14M11 *(b)*	14M12 *(d)*
14M13 *(a)*	14M14 *(c)*	14M15 *(d)*	14M16 *(a)*
14M17 *(b)*			

Chapter 15

Exercises (Section 15.6)

1 *(a)* 816 Ω, 544 Ω
  (b) 748 Ω, 612 Ω
  (c) 714 Ω, 646 Ω
  (d) 693·6 Ω, 666·4 Ω

2 +6·7%

3 *(a)* 13·6 mA
  (b) 42·9 mA
  (c) 2·9 mA
  (d) 0·65 mA
  (e) 6·8 mA

4 *(a)* 330 kΩ±5%
  (b) 10 kΩ±10%
  (c) 8·2 MΩ±20%
  (d) 2·7 Ω±5%
  (e) 390 Ω±10%

  (f) 2·2 kΩ±10%
  (g) 3·9 MΩ±20%
  (h) 39 Ω±2%
  (i) 47 kΩ±5%
  (j) 5·6 kΩ±30%

Multiple-choice answers (Section 15.7)

15M1 (*b*)	15M2 (*d*)	15M3 (*a*)	15M4 (*d*)
15M5 (*c*)	15M6 (*a*)	15M7 (*b*)	15M8 (*d*)
15M9 (*a*)	15M10 (*a*)	15M11 (*c*)	15M12 (*c*)
15M13 (*c*)	15M14 (*b*)	15M15 (*d*)	15M16 (*a*)
15M17 (*b*)	15M18 (*c*)	15M19 (*c*)	15M20 (*d*)
15M21 (*b*)			

Index

voltage being less than the EMF on discharge. The internal voltage drop, given by multiplying internal resistance by current, must be subtracted from the EMF to find the output voltage; that is

$$E - IR_c = U$$

where E = cell EMF, V
  I = current taken from the cell, A
  R_c = cell internal resistance, Ω
  U = cell terminal PD, V

The internal resistance of a cell usually increases with its age, and with ill treatment such as excessive discharging current, standing in the discharged condition etc.

Example 8.1

A cell has an internal resistance of $0 \cdot 02$ Ω and an EMF of $2 \cdot 2$ V. What is its terminal PD if it delivers (*a*) 1 A (*b*) 10 A or (*c*) 50 A?

(*a*) $$U = E - IR_c = 2 \cdot 2 - (1 \times 0 \cdot 02) \text{ volts}$$

$$= 2 \cdot 2 - 0 \cdot 02 \text{ volts}$$

$$= 2 \cdot 18 \text{ V}$$

(*b*) $$U = E - IR_c = 2 \cdot 2 - (10 \times 0 \cdot 02) \text{ volts}$$

$$= 2 \cdot 2 - 0 \cdot 2 \text{ volts}$$

$$= 2 \text{ V}$$

(*c*) $$U = E - IR_c = 2 \cdot 2 - (50 \times 0 \cdot 02) \text{ volts}$$

$$= 2 \cdot 2 - 1 \cdot 0 \text{ volts}$$

$$= 1 \cdot 2 \text{ V}$$

This example illustrates how the terminal PD falls off as the current increases. The effect is noticeable when a car is started at night. The heavy current taken by the starter motor reduces the battery voltage and the lights dim for as long as the starter is used.

The internal resistance of a cell depends on its design, construction, age and condition. Internal resistance can be measured using a high-resistance voltmeter and an ammeter connected with a switch and resistor (Figure 8.9). When the switch is open, no current is taken from the cell (if we neglect the voltmeter current), so the voltmeter reads cell EMF. If the switch is closed, the terminal voltage and current are measured.

Since $$U = E - IR_c$$

$$R_c = \frac{E - U}{I}$$

#200   04-07-2015 12:10PM
Item(s) checked out to MAZON, REBECCA.

TITLE: Inventing beauty : a history of t
BARCODE: 31232001526526269skok
DUE DATE: 05-05-15

TITLE: The artificial face : a history o
BARCODE: 39094006670708893skok
DUE DATE: 05-05-15

$1/day late fee for DVDs & videogames
$1 de multa por dia cada DVD/juego tarde